Six Sigma in the Pharmaceutical Industry

Understanding, Reducing, and Controlling Variation in Pharmaceuticals and Biologics

Six Sigma in the Pharmaceutical Industry

Understanding, Reducing, and Controlling Variation in Pharmaceuticals and Biologics

Brian K. Nunnally
John S. McConnell

CRC Press
Taylor & Francis Group
Boca Raton London New York

CRC Press is an imprint of the
Taylor & Francis Group, an **informa** business

CRC Press
Taylor & Francis Group
6000 Broken Sound Parkway NW, Suite 300
Boca Raton, FL 33487-2742

© 2007 by Taylor & Francis Group, LLC
CRC Press is an imprint of Taylor & Francis Group, an Informa business

International Standard Book Number-10: 1-4200-5439-2 (Softcover)
International Standard Book Number-13: 978-1-4200-5439-2 (Softcover)

Library of Congress Cataloging-in-Publication Data

Nunnally, Brian K.
    Six sigma in the pharmaceutical industry : understanding, reducing, and
controlling variation in pharmaceuticals and biologics / Brian K. Nunnally and
John S. McConnell.
        p. ; cm.
    Includes bibliographical references and index.
    ISBN-13: 978-1-4200-5439-2 (softcover : alk. paper)
    ISBN-10: 1-4200-5439-2 (softcover : alk. paper)
    1. Drugs--Quality control. 2. Six sigma (Quality control standard) 3.
Pharmaceutical industry. I. McConnell, John S. II. Title.
    [DNLM: 1. Drug Industry. 2. Technology, Pharmaceutical--methods. 3.
Efficiency, Organizational. 4. Quality Control. QV 736 N972s 2007]

RS192.N862 2007
338.4'76151--dc22   2007003468

Dedication

For:
Danielle, James, Timothy, Brian,
Annabelle, Jacob, Darcy, and Jonah

Table of Contents

Preface

The world is too big for us. There is too much doing, too many crimes, casualties, violence, and excitements. Try as you will, you will get behind the race in spite of yourself. It is an incessant strain to keep pace and still you lose ground. Science empties its discoveries on you so fast that you stagger beneath them in hopeless bewilderment. The political world witnesses new scenes so rapidly that you are out of breath trying to keep up with them. Everything is high pressure. Human nature cannot endure much more.

Atlantic Journal, **1837**

If you can't stand the heat, get out of the kitchen.

President Harry S. Truman

This book was created to fill a void. We could find nothing that focused on the basic building blocks of understanding and reducing variation generally, and Six Sigma in particular, that was specific to the pharmaceutical industry.

The pharmaceutical industry needs to find new and innovative approaches for everything it does. Discovery and development costs remain high, as do the costs of clinical trials. These costs are subject to rampant inflation, despite efforts by companies to do more with less. Manufacturing, an area that should be able to improve performance continually after each line is commissioned, is under increasing pressure to reduce costs while making quantum leaps in regulatory compliance. Social demands to reduce health costs have placed great stress on the industry. Many pharmaceutical industry executives are convinced that the methodologies that have served the industry to date are now failing, and that they need to be replaced with more promising approaches. The industry might not be a burning platform, but most people working in it smell smoke.

We have not written this book from a purely academic perspective. Both authors have used the approaches and techniques explained in these pages and enjoyed significant success. We have seen deviations and analytical error halved and production doubled. A metamorphosis is possible, but such transformation will not necessarily come easily, especially in large corporations whose emotional and intellectual investment in established methodologies is sometimes considerable.

The conceptual framework contained in this book is much more than a collection of tools and techniques; it provides the basis of a complete operating philosophy. It is this set of mental models that is of greatest importance. If the industry knows what it must do, and why, it will find methods by which to achieve its aims. This is why of the 18 chapters, 13 of them concern core concepts rather than tools and techniques.

We have chosen not to pursue many of the structural elements of Six Sigma in any detail, such as the Define, Measure, Analyze, Improve, Control approach. Tables for calculating the sigma level and methodologies to calculate defects per million opportunities are glossed over. This decision was made not because we thought these elements to be unimportant, but because there were many other references available that cover these subjects more comprehensively than space allowed in this book. In addition, we have placed a focus on the improvement of existing products and processes because that is what people in the industry are asking for. However, it is worth noting that in the long term, the discovery, development, and design areas are more critical.

Case studies are presented throughout the book, along with charts to illustrate these case studies. In some cases, the data have been modified, for example, by inflating or deflating all data by a multiple of (for example) 1.537. In other cases, the scale has been omitted to protect the confidentiality of the data. On occasions, case studies have been simplified for clarity. Nevertheless, the lessons drawn from these case studies are the same as those drawn from the original narratives and data.

It is a shame that the statistical fraternity that pioneered the quality movement did not standardize the symbols used in statistics a long time ago. For instance, in different books the symbols $\bar{x}'$ μ, and η are all used to signify the average (population parameter). This variety can confuse beginners. We have used the symbols used by the pioneering American statisticians, Grant and Leavenworth, because their book, *Statistical Quality Control,* is our principal statistical reference.

Primarily, this book has been written for people working in the pharmaceutical industry. It would please us greatly if it satisfied other needs and purposes, but the intent was to provide people in the industry with the fundamentals, as well as sufficient detail to make a worthwhile start.

It was Thomas Edison who said, "There's always a better way; find it." The pharmaceutical industry could hardly hope for a better role model. Good quality and excellent precision are not destinations. They are a journey. See you on the road.

John S. McConnell
Brian K. Nunnally

The Authors

Brian K. Nunnally, Ph.D., is currently an associate director in Analytical Development for a major pharmaceutical company. He has edited two books, authored many publications in the field of analytical chemistry, and has given numerous lectures in a variety of analytical chemistry subjects. He has applied his knowledge of protein chemistry and analytical chemistry to solve a variety of issues and to make improvements in the analytical measurements made in QC and Development. He is an assistant editor of the journal *Analytical Letters* and a member of two USP Expert Committees (Reference Standards and B&B: Vaccine and Virology). He graduated with two B.S. degrees (1994) from South Carolina Honors College, University of South Carolina, Columbia, and a Ph.D. (1998) in chemistry from Duke University, Durham, North Carolina. He has conducted short courses in statistical process control and is considered an excellent instructor.

John S. McConnell left the Australian Armed Forces in 1982 after more than 13 years service (Royal Australian Engineers). He became Queensland Manager of Enterprise Australia, a non-profit company that had as one of its key objectives the teaching of Dr. Deming's approach to management. In 1987, John founded Wysowl Pty. Ltd., a consulting business he still owns and manages. This is his fifth book. John's greatest strength is his ability to explain issues such as statistics and systems theory in such a way that people from the shop floor to the boardroom are able to understand and use the concepts. His use of metaphor and parables is especially helpful in creating not only understanding, but also impact. He is a gifted speaker and is in high demand on both sides of the Pacific.

1 The Enormous Initial Mistake

Some Thoughts on the Change Process

The change process is seldom easy. It is almost never rational. Always, significant resistance can be anticipated whenever a new approach must displace established practices or beliefs.

The bald, middle-aged man rushed from one woman to another, muttering incoherently and handing them circulars as they traversed the crowded streets of Budapest. The circulars were addressed to prospective mothers and warned them that they should beware of doctors who would kill them. "They say I am mad," the circular continued, "but I seek only to help you. I am your friend, Ignaz Philipp Semmelweis." [1]

Semmelweis was committed. Within a few weeks of his incarceration, the man who discovered not only how infection was transmitted, but also how to prevent it, was dead. Through a combination of accident, analysis, and deduction, Semmelweis had discovered that doctors were unwittingly killing their patients. Failure to properly wash and disinfect their hands before leaving the dissection room in the morgue to attend to their patients spread infection and death. The commonly used term to describe puerperal fever, the cause of these deaths, was "childbed fever." Before Pasteur discovered germs, and before Lister had demonstrated the successful use of antiseptic, Semmelweis was saving lives by preventing infection.

Semmelweis earned his medical degree in 1844. He obtained his Master's degree in Midwifery and was appointed to assist Johann Kline, Professor of Midwifery for Vienna's General Hospital in 1847.

The Lying-in Hospital where Kline and Semmelweis worked had two obstetrics wards. The First Division was a charity ward where patients were usually attended to by student doctors. In the Second Division, wealthier patients were served by trained nurses and midwives. Several times a day a bell tolled in one of the two obstetrics wards of the Lying-in Hospital. The bell announced the death of yet another patient. A sensitive young man, Semmelweis was horrified by the appalling death rate, which was six times higher in his First Division than it was in the Second Division. In some hospitals, the death rate from childbed fever was as high as 26%.

One by one, Semmelweis closely examined and discarded all the published and otherwise listed causes of childbed fever. Included were wounded modesty, guilt and fear complexes (most of the mothers-to-be were poor and many of the babies were illegitimate), sudden variance in weather and temperature, and even cosmic influences were among the fantastic, imagined causes of childbed fever.

Semmelweis was scientifically as well as medically trained. He was competent at statistical analysis and what we now call the scientific method. He discarded most of the proffered causes of childbed fever because they applied equally to the First and Second Divisions. The only significant difference he could find was that the women in the First Division were treated by student doctors and those in the Second Division were treated by trained midwives. As Semmelweis struggled to find a reason for the difference in the death rates between the two wards, serendipity intervened. A vital clue emerged when a friend and fellow doctor, Professor Kolletschka, died of symptoms remarkably similar to those that daily carried away Semmelweis' own patients. Kolletschka had been conducting a postmortem examination and accidentally cut his finger. Within a few days, he was dead.

With no knowledge of germs or infection, Semmelweis had little to guide him. Yet, by skillful analysis and deduction, he correctly concluded that Kolletschka died from some sort of infection contracted from the corpse and that students who regularly dissected corpses as part of their studies were treating his patients. He watched the students conducting their dissections; he saw their hands, dripping with the pus and fluids of putrefied cadavers; he saw those same hands, inadequately cleansed, treating the patients of the First Division, and he concluded that their unhygienic procedures were transmitting infection to the women. Given the complete lack of knowledge of germs or infection that existed in 1847 and that Semmelweis had only deduction to guide him, it was a stroke of brilliance.

Semmelweis insisted that all students and doctors wash their hands in chlorinated lime [$Ca(OCl)_2$] water before entering the ward. Immediately, the death rate in the First Division began to drop. Before long, it had fallen to the same level as the Second Division where midwives treated the patients. In Semmelweis' own words:

> In order to destroy the cadaveric material adhering to the hands, I began about the middle of May, 1847, to employ chlorina liquida with which every student was required to wash his hands before making an examination. After a short time a solution of chlorinated lime was substituted because it was not so expensive. In the month of May, 1847, the mortality in the first Clinic still amounted to over 12 percent, with the remaining seven months it was reduced in very remarkable degree. In the first seven months mortality was 3 percent compared to 11.4 percent prior to introduction of antisepsis. This compared to 2.7 percent in the Second Division. In 1848 the mortality fell to 1.27 percent versus 1.3 percent in the Second Division. In 1848 there were two months, March and August, in which not one single death occurred among the patients of the First Division.[2]

Many doctors, including Semmelweis' superior, Professor Kline, were frustrated and angered by his constant criticism of hospital administration and medical procedure. Kline believed the high incidence of childbed fever was due to the hospital's ventilation system, an idea that fitted the then-popular miasmatic theory of disease. Semmelweis' insistence that even Kline wash his hands before entering the ward angered his superior to the point where Kline eventually refused. Semmelweis continued to harangue Kline and anyone else displaying doubt about his theory. His data and case studies were compelling, but many senior doctors, including Kline, were unable to accept that their own lack of hygiene killed patients. When Semmelweis' appointment expired in 1849, Kline extracted his revenge and refused to renew it. Semmelweis appealed, triggering a faculty feud between Klein and a Semmelweis supporter, Professor Skoda. Klein won. Angry and frustrated, Semmelweis left for Budapest in 1850.

Soon after his return to Budapest, Semmelweis was made head of the obstetrical service at the St. Rochus Hospital in Pest where he conducted a six-year clinical trial of his theory and achieved a mortality rate of 0.85%. In 1855, his academic ambitions were fulfilled by his appointment as Professor of Midwifery at the University of Pest. The obstetrical service he took over was a disaster. In the first year of his tenure, he drove the puerperal fever death rate to 0.39%, an almost unbelievable record. In 1861, he published a book, *The Etiology, Concept, and Prophylaxis of Childbed Fever.*

Semmelweis' and his theory were attacked by the self-serving Professors of Midwifery who formed the medical hierarchy. The death rates from childbed fever for these esteemed and powerful men ranged as high as a cruel 26%. The statistical evidence suggesting these eminent professors were angels of death only served to strengthen their opposition to Semmelweis. Semmelweis' criticism of anyone who denied his doctrine in spite of the high mortality rates in their own institutions was direct and incisive. In an open letter to Professor Scanzoni of Würzburg who had demeaned his work, he wrote:

> Your teaching (that the Würzburg epidemic of childbed fever is caused by unknown atmospheric influences or puerperal miasma) is false, and is based on the dead bodies of lying-in women slaughtered through ignorance...I have formed the unshakable resolution to put an end to this murderous work as far as lies in my power so to do...(If you continue teaching your students this false doctrine), I denounce you before God and the world as a murderer, and the History of Puerperal Fever will not do you an injustice when, for the service of having been the first to oppose my life-saving doctrine, it perpetuates your name as a medical Nero.[2]

The years of denigration, frustration, and angst took their toll. Semmelweis had a nervous breakdown and was relieved of his post. In addition, his recently published book was widely derided and criticized. His condition worsened and he was admitted to an asylum. His death is something of a mystery. Some sources have him dying in the ironic manner of a Greek tragedy, deliberately infecting himself with a scalpel during an autopsy.[1] Another version has him accidentally cutting his finger during an operation immediately before his commitment.[3] Still others claim he died at the hands of asylum staff. Whichever is true, it was a cruel fate for the father of infection control.

WHY?

What went wrong? Why did Semmelweis find it so difficult to convince others in his profession that his approach to hygiene worked well and saved lives? Semmelweis had a lot going for him. Certainly, he had good data. Everywhere he went he reduced deaths to amazingly low levels. It would have been difficult to argue with his data, or his compelling case studies. Perhaps our understanding can be improved by an extract from a paper by Dr. Myron Tribus, *The Germ Theory of Management.*[4] In part, Tribus' paper reads:

> Doctors administer to the needs of their patients according to what they learn in school and in their training. They also learn from experience. They can only apply what they know and believe. They have no choice. They cannot apply what they do not know or what they disbelieve. What they do is always interpreted in terms of what they understand is, "the way things work." As professionals they find it difficult to stray too far from the common knowledge and understanding of their profession. They are under pressure to follow 'accepted practice'. In this regard, Doctors are no worse than the rest of us. We are all prisoners of our upbringing, our culture and the state of knowledge of our teachers, mentors and fellow practitioners. [4]

Learning is incidental. One does not always have to work at it. We learn much by simply being aware of the world around us. Unfortunately, it is as easy to learn superstition as it is to learn knowledge. Any teacher of adults will attest that one of the greatest barriers to learning is not ignorance; it is the illusion of knowledge or the presence of superstitious knowledge. Those students who have no pre-existing beliefs to displace learn faster than those who do. Our children learn that the world is round with much greater ease than did many scholars in the time of Galileo because they have nothing to unlearn or reconcile at the outset.

Suppose we return to the year 1847, and to the Vienna General Hospital. Also, imagine that you have been working alongside Dr. Semmelweis and were jointly responsible for the remarkable results discussed earlier. The years of toil have been rewarded with a new approach that demonstrably saves many lives, and when spread across the hospitals of Europe will save millions more.

Your most obvious task is to spread this new knowledge, and Professor Skoda has arranged a seminar where your paper on the subject will be presented to several hundred of the most eminent medical men of Europe. They come from universities, hospitals, and private practice. Among them are the intellectuals whose books and papers on midwifery are compulsory reading for students. At this very moment these men sit and listen to Professor Skoda introduce you to the audience. From your position behind the curtain, you watch the best medical minds Europe has to offer wait for you to enter and present your paper. In a few seconds, your job is to step onto the stage and convince these men that they are killing their patients. You have a moral and ethical responsibility to explain to them that because they do not follow the strict hygiene practices detailed in your notes, they are spreading infection every time they touch a patient or an instrument. Your task is to persuade them to discard much of their accumulated knowledge and experience and start anew.

How do you feel?

Now, let us turn the tables. Imagine you are a wealthy, successful doctor. You are the superintendent of a large hospital, and are held in high regard not only by your fellow doctors, but also by society. Your advice is sought widely. At the local university, students hang on every word during your weekly lecture. How do you think you will react to Semmelweis and his associate? How will you feel if the journalist sitting in the wings publishes the notion that you and your practices are a threat to every patient you touch?

How did those who were confronted by the theories of Copernicus and Galileo feel? How did the physicists feel when Einstein shattered their cherished notions of absolute time and space? Some scientists were so incensed at Einstein's theory that they wrote a rebuttal, *100 Authors against Einstein.*[5] Of all people, would we not expect scientists to understand that relativity theory was a better way of describing the universe and see the huge potential it offered? Why is it that whenever a breakthrough idea presents itself, we humans have so much difficulty understanding and accepting it (at least initially; a generation or two later it is easy)?

Semmelweis' story led to the creation of the term "Semmelweis Reflex," meaning the dismissing or rejecting of new ideas, without thought, inspection, or experiment. It also refers to the "mob" mentality that flows from "groupthink" and the refusal to consider a new point of view.

THE ULTIMATE CURSE

From 1950 onward, the Japanese economy grew until it was a serious threat to the survival of several industries in the U.S. During the 1970s and 1980s, hundreds of study tours went to Japan in search of the quality and productivity "secrets" of Japanese companies. The great majority returned to the U.S. with only part of the answer, or with conclusions that were irrelevant, because the wrong questions had been asked. The visiting Americans were so impressed with the Japanese workforce that many assumed Japan's "secret" was a combination of her culture and a dedicated and productive workforce. They were as sure of this as the doctors' in Semmelweis' day were of the miasmic theory of disease. Also, the American managers were sure that the management theory they were using was correct, despite the fact that the Japanese did not use much of it.

As America's economy continued to struggle, men such as Dr, W.E. Deming, Dr. J. Juran, and Dr. H. Sarasohn were largely ignored in America, and yet it was primarily these men, according to the Japanese, who were responsible for Japan's rise to the status of an economic superpower.[7] For many years, these men could not make themselves heard in America. The scene was set for yet another "Semmelweis Reflex," and that's exactly what happened, and is still happening in many places. Similar to Semmelweis, their logic, data, and case studies were largely ingored.

The ultimate curse is to be a passenger on an ocean liner, to know that the ship is going to sink, to know precisely what do do to prevent it, and to realize that no one will listen.[6] This is the curse visited on men such as Deming, Juran, and Sarasohn for the last fifty years.

Now we see people in the pharmaceutical industry trying to convince the industry that the approaches developed by Shewhart, Deming, Smith, Juran, Harry, and others hold the promise of improved quality and productivity as well as fewer deviations and regulatory issues.

Already, some of these change agents have experienced their own Semmelweis Reflex. One man, a highly regarded Ph.D., led a thrust to reduce analytical error in a pharmaceutical laboratory. Under his guidance, one of his laboratory statisticians conducted a project to stabilize and reduce variation for a problematic laboratory test. The project was spectacularly successful. Test error plummeted. They were prevailed upon to present the results of this project to colleagues and peers, with a view to spreading the methodology. The audience was almost as unresponsive as were the physicians hearing Semmelweis' case studies for the first time. They could not see a problem. They were meeting the required standards. Even if a similar project in their laboratories might yield similar results, why should they bother to drive analytical error to even lower levels?

No argument moved the detractors. Neither improved service to customer departments nor the potential to reduce regulatory deviations impressed them. At the time of writing, the status quo remains. It is likely to remain in place until senior management removes options to conquering variation by making it a strategic imperative.

A METAMORPHOSIS IS POSSIBLE

In order to illustrate what is possible, we present a chart from another operation (Figure 1.1). It shows laboratory controls plotted as a control chart for the period immediately before and after the drive to conquer test error was made. Readers are free to draw their own conclusions, but the authors feel confident that Dr. Semmelweis would have approved of the work done.

THE ENORMOUS INITIAL MISTAKE

Schopenhauer wrote:

> Almost without exception, philosophers have placed the essence of mind in thought and consciousness; man was the knowing animal, the animal rationale. This ancient and universal radical error, this enormous initial mistake, must before everything be set aside.[8]

Six Sigma is rooted in the rational world of statistics, but the world of business is not a wholly rational place.[9] Business is about people, so Six Sigma is about people. People are not always rational. Each of us leads an emotional existence. Semmelweis tried to convince others with an argument based on fact, data, and logic. He failed. Leaders and change agents would be well advised to take into account the emotional aspects of change.

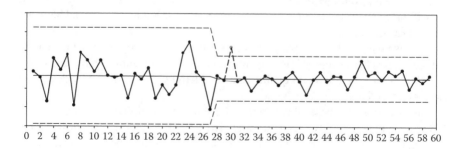

FIGURE 1.1 Laboratory controls as a Shewhart control chart.

Only a storm of hot passion can turn the destinies of peoples, and he alone can arouse passion who bears it within himself.

It alone gives its chosen one the words which like hammer blows can open the gates to the heart of a people.[10]

So wrote Adolf Hitler in *Mein Kampf*. Hitler was a monster. However, it would be a mistake to allow his evil mind to blind us to the fact that he was one of the most successful transformational leaders in history. When he came to power, Germany was an economic mess. Inflation was rampant; six million people were unemployed and national morale was at a low ebb. Hitler transformed Germany and its economy. He may have been a monster, but he understood well that emotional impact was an essential ingredient for rapid and profound change.

Many business leaders face similar issues as they prepare to transform their businesses with a Six Sigma approach. A clear lesson from the history of Six Sigma is that emotional impact, the withdrawal of options to participating, and alignment of the organizational structures to a Six Sigma philosophy (as opposed to a bureaucratic ticket punching exercise) are essential. In particular, conquering variation to Six Sigma levels must become a primary and non-optional strategic objective for managers and technical people throughout the business. When parts of the industry pursued statistical process control as an improvement methodology, some companies got a lot of pretty charts and not much else. It is the outcome, reducing variation, that is important. One of the strengths of Six Sigma is the establishment of a unifying objective for everyone in the business. Leaders and change agents who ignore the vital elements of aligned aims and structures, as well as the creation of emotional impact are in peril of suffering a fate similar to Semmelweis.

ONE POINT LEARNING

1. All organizations contain elements of superstitious learning. For most people, unlearning old beliefs and relinquishing established methods requires emotional impact.
2. Change occurs at three levels: rational (new knowledge), physical (new physical structures, including information, financial, pay and promotion systems, as well as organizational structures), and emotional (new value systems and the creation of emotional impact). Leaders and change agents must address all three.[9]
3. In particular, having the achievement of Six Sigma levels of performance as a non-optional and organization-wide objective is vital.

REFERENCES

1. Historical feature, *Daily Mirror*, Sydney, November 2, 1989.
2. W.J. Sinclair, *Semmelweis, His Life and His Doctrine*, Manchester University Press, Manchester, 1909.

3. J.L. Wilson, *Stanford University School of Medicine and the Predecessor Schools: An Historical Perspective,* 1999, Lane Medical Library Digital Document Repository, http://elane.stanford.edu/wilson/
4. M. Tribus, *The Germ Theory of Management,* SPC Press, Knoxville, TN, 1992.
5. A. von Brunn, 100 Authors against Einstein (1931), in *Physics and National Socialism; An Anthology of Primary Sources,* K. Hentschel, Ed., Birkhäuser, Boston, MA, 1996.
6. M. Tribus, *Quality First,* IQMI, Sydney, 1988.
7. K. Ishikawa, *QC Circle Koryo,* JUSE, Tokyo, 1980.
8. W. Durant, *The Story of Philosophy,* Touchstone, New York, 1926.
9. W. Scherkenbach, *Deming's Road to Continual Improvement,* SPC Press, Knoxville, TN, 1991.
10. A. Hitler, *Mein Kampf,* Pimlico, London, 1992.

2 The Origins of Six Sigma

From Where Does the Term Six Sigma Spring?

In this chapter, the reader is taken on a guided tour of the origins and development of the Six Sigma concept. As will be seen, its origins can be traced back to the early part of the last century.

GENESIS

The genesis of Six Sigma can be traced to Motorola in 1979 when executive Art Sundry stood up at a meeting of company officers and proclaimed, "The real problem at Motorola is that our quality stinks!"[1] Some time later, Bill Smith, an engineer at Motorola's Communications Sector, was studying the correlation between a product's field performance and the variation and rework rate in manufacturing. In 1985, Smith presented a paper that concluded that products produced in processes that had a significant rework rate had higher field failure rates than those produced in processes that had low variation and negligible rework rates.

At about the same time, Dr. Mikel Harry, at Motorola's Government Electronics Group, created a detailed and structured approach involving statistical analysis to improve product design and manufacturing performance, thereby simultaneously improving the product and reducing costs. These approaches developed into what is now known as Six Sigma.

As this chapter was researched, many descriptions of Six Sigma were found. One of the more comprehensive was:

Six Sigma is an approach to conquering variation that was developed by Motorola in the early 1980s. There are two definitions of Six Sigma, both of which are appropriate, a technical definition and a cultural definition, as follows:

1. **Technical Definition.** Six Sigma is a statistical term used to measure the performance of products and processes against customer requirements. By definition, a step in the process that is operating at a Six Sigma level produces only 3.4 defects per million opportunities.
2. **Cultural Definition.** Six Sigma is a management philosophy and a cultural belief system that drives the organization toward world-class business performance and customer satisfaction. It is based on scientific principles, including a decision-making process based on fact and data.

The Six Sigma approach aims to drive defects and "things gone wrong" to extraordinarily low levels, to increase first pass yield and to consistently exceed customer expectations. First pass yield is a measure of the percentage of jobs that exit the process right, on time, first time.

While this description is essentially correct, it fails to explain satisfactorily the conceptual core of Six Sigma, so that is where this chapter will begin.

UNDERSTANDING AND REDUCING VARIATION

FROM WHERE DOES THE TERM SIX SIGMA SPRING?

Sigma is the Greek letter that represents standard deviation. It can be calculated for any data set. However, when teaching or guiding laymen, it is best to start with a geometrical description. Note that on either side of the average for a normal or bell-shaped distribution, there are in fact two curves. One is concave and the other is convex. If these curves are continued, as shown as dashed lines in Figure 2.1, there is a point of inflection where the two curves touch. If a line is dropped from this point of inflection to the baseline, the distance between this vertical line and the average is one standard deviation.[2]

Imagine parts are being made for a medical devices assembly operation in the pharmaceutical industry. As the parts are manufactured, the critical variables are measured and these data are plotted to make frequency distributions. For the purpose of illustration, assume that the distributions created only just fit into the customer's specifications, as illustrated in Figure 2.2. If the processes were improved so that the degree of variation for this particular variable halved, the specifications would now be Six Sigma from the center line. The specifications remain unchanged, as does the distance between the center line and the specifications. However, because the degree of variation in the data has been halved, so too has the standard deviation. This is only true of bell-shaped distributions, but the relationship expressed is adequate for illustrative purposes.

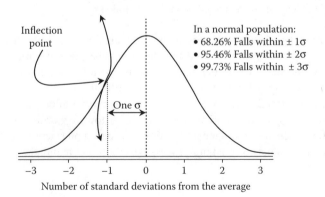

FIGURE 2.1 The normal curve and a geometric representation of sigma.

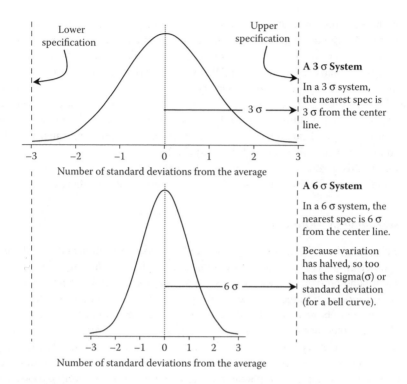

FIGURE 2.2 Three- and six-sigma systems compared (assuming stability).

This, then, is the origin of the term Six Sigma.[3] In a Six Sigma system, every part and every processing step shows variation such that the nearest specification is 6 away. If the distribution is correctly centered and bell-shaped, it must display no more than half the variation allowed by the specifications to be a Six Sigma system. Nevertheless, many people are inclined to ask why one would bother to reduce variation so far below the specified requirements.

EARLY SIX SIGMA COMPANIES

Most of the original Six Sigma companies were assemblers. They assembled parts that were sometimes made by external suppliers, sometimes by internal suppliers, and sometimes by a mixture of both. Motor vehicles, cell phones, and other electronic items are a few well-publicized examples. It was from this environment that the Six Sigma movement initially arose, although it has much wider application. Three of the firms that put the Six Sigma approach in the spotlight were Motorola, Allied Signal, and GE.

Genesis — The Motorola Experience

By the early to mid-1980s, Motorola had sold off elements that once were the core of its business, such as television and Hi-Fi manufacturing. Many of the businesses Motorola sold off were purchased by Japanese companies that made the operations they purchased more profitable with little or no capital expenditure. It was at about this time that Art Sundry made his pronouncement. Some people in Motorola started to look for answers. A few of the Motorola people did a study of all the best companies they could access. They found that the best performing operations (in terms of quality, productivity, warranty repair rates, and customer service levels) were performing at a level about which they dared not dream. For most of the people at Motorola, improved quality meant higher cost. Yet the excellent companies they studied provided good quality at low costs. A closer inspection revealed that the outstanding companies they studied were all Six Sigma operations.[3]

Initially, this puzzled them. It seemed to make no sense. Why go to all that trouble and expense? Why not be satisfied with meeting the specifications?

The Awakening at Motorola

Bill Smith figured out the crux of the issue. It was a mix of variation and complexity. He and other engineers were able to demonstrate mathematically that to get both great quality and productivity was impossible if every part/assembly/component/raw material was delivered to the next step in the process with three-sigma quality. They went to their cell phone assembly process and counted the number of base components. They counted not assemblies, but every piece of wire, every transistor, every contact and switch, every soldered joint. Then they counted the number of processing steps. These counts were combined to arrive at what was called the number of "events" required to make a cell phone. Here an event is defined as every opportunity for something to go wrong, or every opportunity for a defect to occur. The count for a cell phone was about 5000 events. Imagine how many events there are in the assembly of a car or of an airplane. The complexity becomes staggering.[3]

It was observed that the crème de la crème plants had a very high "first pass yield," that is, a high proportion of finished products that were right (met all specifications), on time, first time (no rework or similar fiddling). This made engineering and economic sense. In addition, their data suggested a strong correlation between high first pass yield and low field failure and warranty repair rates. Some at Motorola wondered how they could achieve first pass yield figures over 95% when their combined rework and reject rates were far greater.

What they then did was to create a model, assuming statistical stability in every component and event (a highly unlikely scenario with 5000 events), where the variability in every event was at three sigma (specifications three sigma from process mean). By definition, a three-sigma event produces 99.73% inside specs, assuming normality. The model showed that after about only 1000 events, the first pass yield fell to zero. At that level of variation, all event supervisors or managers could claim that they were doing their job; their three-sigma processes were producing out-of-spec components/assembly at a rate of only 0.27%. However, even if every component and every assembly step had three-sigma variation, the first pass yield was going to be very poor unless there were very few events. Table 2.1 shows how the first pass yield fell as the number of events rose in a three-sigma system. Three-sigma systems worked when manufacturers made muskets and wagon wheels. They even worked reasonably well for early machines. As Deming was fond of

TABLE 2.1
First Pass Yield for Three- and Six-Sigma Systems Compared[a]

Number of Events	Three Sigma System	Six Sigma System
1	99.73	99.99
10	97.33	99.99
20	94.74	99.99
40	89.75	99.99
60	85.03	99.99
80	80.54	99.99
100	76.31	99.99
200	58.23	99.99
400	33.91	99.99
700	15.06	99.99
1000	6.70	99.99
3000	0.00	99.99
10,000	0.00	99.99

[a]Assuming stability.

saying, "we no longer live in that world." Note the enormous increase in yield for a Six Sigma system.

One can enter any multiple of sigma for each event to try to mimic a certain process. Most of Motorola's processes were operating at approximately four sigma. The factorial model was not a good fit for Motorola's factories. Some processes had even worse first pass yield than the model predicted. This was expected, at least in part, because perfect stability was impossible. Incoming raw material will show lot-to-lot variation. Temperature will vary. Lab analysts will grow tired and make mistakes. Machines will drift out of adjustment. Tool steel will dull. Bearings will wear and vibrate. Refractory will erode. All these things will happen, at least to some degree, in even the best-managed plant. This is why most plants operate at four-sigma levels of quality to provide a buffer against these issues. Bill Smith went one step further with the observation that long-term variation was impacted significantly by shocks to the process that knocked the process average off target, often by as much as 1.5 sigma.

Motorola reworked the model. Keeping the range for each event unchanged, this additional variation was simulated by allowing the process mean for each event to drift back and forth by 1.5 sigma as seen in Figure 2.3.

Here a model is being discussed. It is worth recalling that while all models are wrong, some are useful.[4] This model is useful, but it is wrong. We ought to place no greater significance on the 1.5-sigma shift to the mean than did Bill Smith. It is no more than an attempt to factor in shocks to the system. No factory behaves exactly like these models. Nevertheless, when the Motorola people rechecked their own data and that from the Japanese electronics and car manufacturers, they found that the model was a reasonable fit. In particular, the best of the best plants all had a first

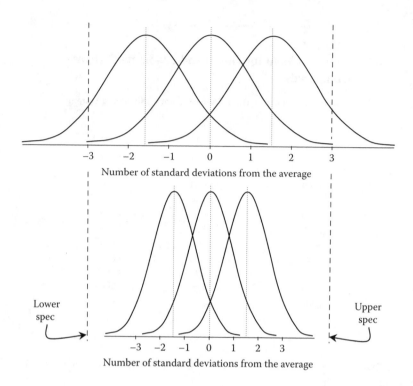

FIGURE 2.3 Three- and six-sigma systems drifting ±1.5 compared.

pass yield that fit the Six Sigma model in Table 2.2 quite well. With every event at Six Sigma, even if the event mean is allowed to drift by ±1.5, first pass yield will be about 96% after 10,000 events, as noted in Table 2.2.

The data in Table 2.2 show how some Japanese operations achieved such seemingly impossible productivity and quality figures simultaneously. What Motorola discovered was that the greater the complexity, or number of events, the greater was the impact of variation. They had to either reduce the number of events, greatly reduce variation, or some combination of both. This is not news for people with an operations research background. It is what Little's Law predicts (see Chapter 8).

Wernher von Braun, architect of the U.S. Space Program, recorded the complexity problem much earlier. He noted that if a large number of components must function for a system to accomplish its objective, the probability of system success diminishes rapidly as the number of components increases, unless the reliability of each is essentially perfect.

Stirrings at Ford

Another group that made a similar discovery in the early 1980s was Ford. Ford's planners realized that the transmission plant that was to make Ford's first U.S.-built front-wheel-drive automatic gearboxes could not make enough transmissions for the

---

TABLE 2.2
First Pass Yield for Three- and Six-Sigma Systems Compared[a]

Number of Events	Three Sigma System	Six Sigma System
1	99.32	99.99
10	50.08	99.99
20	25.08	99.99
40	6.29	99.99
80	0.40	99.97
100	0.10	99.97
200	0	99.93
400	0	99.86
700	0	99.76
1000	0	99.66
3000	0	98.99
10,000	0	96.66

[a]Assuming a drift in mean of ±1.5 sigma.

---

predicted production rates. Ford had recently acquired a sizeable proportion of Mazda, so they asked Mazda to make identical gearboxes and ship them to the U.S. Mazda did this, and it delivered the gearboxes to the U.S., freight paid, for less than they could be shipped out of the U.S.-based Ford plant.

Not long after launch, the dealers and the warranty people at Ford noticed significant differences in performance. The Mazda boxes were quieter, smoother, and had a reduced warranty repair rate when compared to the U.S. manufactured units. Ford dismantled a sample of U.S. boxes and discovered that for the critical components they were running at about four sigma. They were pleased. It looked like an impressive result.

Then they dismantled some Mazda units. In at least one case, they needed to buy a new instrument for the inspector because the one she had could not adequately detect variation between the parts. The components were manufactured so that the piece-to-piece variation was at about six sigma. This staggered them. Again, it made no sense. Who would do that and why?

They went back to Mazda and soon found some of the answers. It surprised many to discover that the Japanese had learned about variability reduction from Americans such as Deming and Juran. In addition, Mazda had determined that assembly costs were a very large element of the total cost of a gearbox. So, they put a lot of effort into the manufacturing of components. As one Ford executive said, the specifications disappeared "over the horizon." The outcome was brilliant first pass yield. Just as importantly, the parts, "practically fell together" and assembly time as well as labor used for assembly plummeted. Simultaneously, work-in-progress fell as did cycle time (as it must, in accordance with Little's Law). The manufactured cost of the components at Mazda was actually higher than it was in the U.S. at Ford, but the much-reduced assembly costs meant a unit that was not only superior, but that also cost less to make overall. The reduced warranty repair rate was an additional bonus, as was the improved customer satisfaction level.[5]

Most manufacturers and assemblers in the U.S., Australia, and Europe are doomed before they start. The most common style of management forces every manager at every step in the process to try to reduce costs within a given department or other organizational component. It can be extraordinarily difficult to get department managers in the first few steps of a process to increase their costs by 1 million dollars per annum so that downstream managers can reduce their costs by 2 or more million dollars per annum. Instead of objectives for, and management of, the entire production process, every step in the process has individual budget targets, and people are held accountable for them. There is a better way, and the Japanese learned it a long time ago from Americans such as Deming and Juran. Six Sigma uses the same process-based approach.

Mazda chose to manage the entire process and to manage total cost. In this case, higher cost components were required so that significant savings could be harvested from the assembly process.

Further Illustration — Vial Capping Issues

At a parenterals plant where vials are filled with a liquid drug, issues with failed caps had plagued the line for some time. A statistical study revealed that the vials, all of which were manufactured in a six-cavity die, met specification easily, as seen in Figure 2.4. However, when the data were stratified by cavity it became apparent that cavity 3 was different from the remainder. Vials from this cavity were responsible for nearly all the failed caps. All vials met specifications, but this is not a Six Sigma system. In fact, none of the cavities taken in isolation is a Six Sigma system. This example is discussed in more detail in Chapter 9.

Understanding and reducing variation lies at the heart of the Six Sigma concept. This imperative can be traced back at least as far as the industrial revolution. In the first half and middle of the last century, great minds such as Shewhart, Deming,

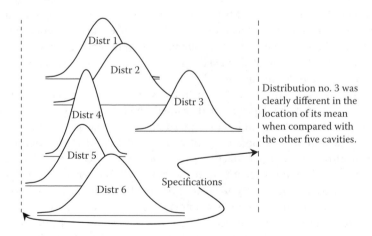

FIGURE 2.4 Distributions for vial necks.

Juran, and Taguchi established the critical role that understanding and reducing variation played in improving quality, productivity, and competitive position. Six Sigma has an excellent ancestry.

What the Six Sigma concept did very well was to explain how some Japanese companies could lay claim to defective rates expressed as three to six parts per million. To those Western managers and technical people who were hard pressed to achieve defective rates of less than 5 to 10%, the claimed Japanese outcomes seemed light years away from their current reality. In many cases, the Japanese figures simply were not believed because the improvement needed to reach these levels of performance in most Western companies seemed several orders of magnitude away from their existing situation. However, the Six Sigma approach gave many something at which they believed they could realistically aim. If a business was currently operating at a quality level of about four sigma, which was where the Motorola studies showed most Western manufacturers to be, it was not too difficult to accept the task of reducing variation at each event by about one-third. This would make the business a Six Sigma business, and the seemingly impossible defective rates being claimed by the Japanese would be achieved in the West.

At Motorola, the variability objective set for every manager was expressed in terms of the first pass yield that would be achieved if every event reached a level of variation described as Six Sigma. This is sometimes referred to as the "shotgun" approach because reducing variation was a task for every person in the business. Motorola then added what we might describe as a "sniper rifle" approach, where advanced statistical techniques were used to identify and target those variables that gave the greatest leverage.

> Statistical thinking will one day be as necessary for efficient citizenship as the ability to read and write.

H. G. Wells, 1925

UNDERSTANDING THE SIGMA LEVEL

Not all processes have both formal specifications and variables that can be measured against those specifications. One example would be the result of inspection of an assembled device. Another would be inspecting the fit of caps on vials. The parts either fit properly or they do not. The cap on the vial is creased or it is not. In the service areas of the pharmaceutical industry, a similar situation is found. Many of the data present as attributes (good/bad, in/out, present/missing, on time/late) or in such a way as to make the estimation of the variability in the process difficult.

In these cases, we can use a complimentary approach. We know that by definition a three sigma level of performance produces three defects per thousand and that a Six Sigma process produces 3.4 defects per million. Using the normal curve formula it is a simple exercise to calculate the level of defects for any given degree of variation, or sigma level, as noted in Table 2.3. The defects per million opportunities (DPMO) figures in Table 2.3 assume that the process mean will either drift or be shocked randomly by ±1.5 .

TABLE 2.3
Determining the Sigma Level[a]

Short-term Process Sigma	DPMO	Short-term Process Sigma	DPMO
0.1	920,000	3.1	54,800
0.2	900,000	3.2	44,600
0.3	880,000	3.3	35,900
0.4	860,000	3.4	28,700
0.5	840,000	3.5	22,700
0.6	810,000	3.6	17,800
0.7	780,000	3.7	13,900
0.8	750,000	3.8	10,700
0.9	720,000	3.9	8190
1.0	690,000	4.0	6210
1.1	650,000	4.1	4660
1.2	610,000	4.2	3460
1.3	570,000	4.3	2550
1.4	540,000	4.4	1860
1.5	500,000	4.5	1350
1.6	460,000	4.6	960
1.7	420,000	4.7	680
1.8	382,000	4.8	480
1.9	344,000	4.9	330
2.0	308,000	5.0	230
2.1	274,000	5.1	150
2.2	242,000	5.2	100
2.3	212,000	5.3	70
2.4	184,000	5.4	40
2.5	158,000	5.5	30
2.6	135,000	5.6	20
2.7	115,000	5.7	10
2.8	96,800	5.8	8
2.9	80,800	5.9	5
3.0	66,800	6.0	3.4

[a] Assuming a drift of ±1.5 sigma in process mean.

If the actual level of defects is measured at each critical to quality (CTQ) event in the process, the DPMO can be calculated at each event using the formula:

$$DPMO = \frac{Total\ Defects \times 1,000,000}{Total\ Opportunities}$$

For example, if during a run of 125,000 vials online inspection revealed 595 creased caps, then the defect rate is calculated as 4760 DPMO. In Table 2.3, the figure closest to this defect rate in the DPMO column is 4660, which corresponds with a sigma level of 4.1.

It is important to understand that the sigma level is calculated for each CTQ event, not for the entire process. The most common form of measurement for the entire process is first pass yield. This is calculated by multiplying the proportion yield for every event in the process together, in serial (as was noted earlier in Table 2.1 and Table 2.2). To provide a simplified example, suppose there are only six major events in a device assembly process. Imagine that the first pass yield for each event, expressed as a proportion, is 0.98, 0.93, 0.88, 0.99, 0.91, and 0.89. The product of these events is 0.64, or a first pass yield of only 64%. Some companies do these calculations on a systematic basis. Others choose not to, and instead focus on the sigma level of each CTQ event, knowing that if the sigma level has been driven to Six Sigma levels, then the first pass yield will be very high for even the most complex process.

GAINING GREATEST LEVERAGE

The Sniper Rifle Element

Manufacturers everywhere know that sometimes no matter how much effort is expended, the limits of existing technology can prevent further variation reduction. It can be difficult, expensive, or impossible to achieve Six Sigma levels of variation for some events in the process. One example came from the early Motorola experiences. One component they struggled with was a crystal used in the manufacture of a cell phone. Crystals are not manufactured, they are grown, and there was significant variation from crystal to crystal. Initially, Motorola had been laser trimming these crystals to achieve the desired uniformity in outcomes. The approach worked, but it was very expensive. Using some sophisticated multi-variable statistical analyses, Motorola discovered that if two of the other variables involved could be held to extremely tight tolerances, then the interaction between these variables and the crystals was such that the crystal could be allowed to vary significantly without loss of performance. Instead of reducing the variation in the crystals, they discovered a much cheaper way to achieve the desired outcomes.[3]

This was the "sniper approach" element of Six Sigma developed at Motorola. It involved projects using advanced statistical methods that helped to locate those variables that, if addressed, give the greatest advantage in the drive toward improving processes and products. There are many sophisticated statistical methods that one can use during the conduct of such projects,[4] but these are the preserve of highly-trained people, and they are beyond the scope of this book. Instead, an overview of some of the principles associated with Little's Law will be provided.

Lessons from Little's Law

Little's Law states that output volume is a function of cycle time, which in turn is a function of variation.[6] If variation in the flow of material/parts through the process is successfully reduced, volume output will rise provided work in progress remains constant. The Dice Experiment (Chapter 8) illustrates this law in action. However, in the Dice Experiment, every event has the same capability and the same stable variation in terms of volume throughput. Any manufacturer knows that real factories look quite different. Some events are more critical than others. However, Little's Law offers insights that are generally very helpful.

Both the Motorola model and the Dice Experiment show that leverage is gained by working upstream in the process; by working on the inputs and the first few events. If variation is reduced in the first few events, there will be a flow effect through the remainder of the process. However, in most factories that is not where the improvement efforts are focused. Any process has what we might call a "transformational event." This is the event or step in the process where the alchemic transformation takes place. In the service elements of our businesses, "transformation" might be the movement of a product in the distribution system.

Too often, the most talented people in a business can be found intensely studying the transformational event when the root causes of most problems are to be found earlier in the process. A good rule of thumb is that for all the issues we see at the transformational event, 50 to 80% of the root causes of these problems will be found earlier in the process, up to and including set-up of the transformational event. Many of the issues that are not solved by addressing the early events have their causal relationships exposed after the variation in the early events has been conquered. Statistical methods are available to find those variables that offer the greatest leverage. Another approach is to start every project by focusing on the inputs and early steps of the process. Both approaches have their merits. They both work. This is not an either/or situation. Rather, it is one more properly characterized as an "and" situation.

There are many simple tools and techniques that we can use in project work to isolate key variables. The various fishbone or Ishikawa diagrams, flow diagrams, frequency distributions, Pareto charts, and simple two-variable correlations are some examples. There is a place for sophisticated statistical techniques, but equally there is a place for the simpler approaches that everyone can master. A search for the key variables that provide greatest leverage need not always involve heavy-duty statistics.

The empirical evidence is overpowering. Companies that achieve Six Sigma levels of performance have higher revenues and lower costs than those that do not. There is a price to pay for any level of variation above near perfection.

DESIGN

Many companies have enjoyed significant success in improving existing products, services, and processes, only to discover that every new product or process brings a new batch of problems to be solved. Despite many years of debate and discussion about robust design, reengineering, design for manufacturability and maintenance, and more recently design for Six Sigma (DFSS), some companies have yet to create a culture and an approach that designs potential issues out of the manufacturing process before construction of the facility, as well as optimizing performance in the customer's hands. One significant hurdle is the presence of barriers between functional areas such as research or discovery, product design, development, facility design and construction, commissioning, and operations. There must be a strategy to overcome these issues. They will not disappear just because they have been identified.

Some businesses create a design team immediately when a promising product emerges from the research or discovery phase. These teams are compromised of people from development, facility design and construction, compliance, and production. Their job is to design problems out of the remainder of the process, from development to full production, and to have as many activities as possible happening in parallel rather than in series. Some structural elements that help to remind people everywhere that this process is taken seriously are necessary. One is to transfer-price the costs associated with poor design, development, and facility design and construction back to the relevant areas. This exposes such costs so they are not hidden in the construction and production costs, an all too common occurrence in the pharmaceutical industry.

The literature generally agrees that about half of the costs associated with waste, rework, and non-compliance can be traced to product design, development, and process design. Even businesses that are successful at revolutionizing performance of existing processes and products fall short of optimum performance because of limitations in design. In addition, every new product and process brings new problems to be addressed. Compare this with new facilities that are stable and running at full production within weeks of commissioning. The potential is enormous.

SUMMARY

The "sniper rifle" and design elements of the Six Sigma approach involve locating and addressing those variables that, if successfully addressed, give the greatest leverage. Sometimes simple tools will do the job. Sometimes sophisticated statistical modeling or tools are called for. Because of the size limitations of this book, the more advanced techniques will not be addressed here. Nevertheless, if readers are aware that some very powerful statistical tools are available and know who to turn to when their use is called for, advancement can be made.

SOME STRUCTURAL ELEMENTS OF SIX SIGMA

A BUSINESS STRATEGY

Six Sigma is not only a quality initiative. It is also a business strategy. [1] Unless the approach is part of the mental models of senior management and is being driven by them, it will sub-optimize at best and fail at worst. Six Sigma brings focus onto the cost of quality. In the pharmaceutical industry, the cost of compliance needs to be added to this. The irony here is that the cost of preventing deviations is often tiny compared to the cost of raising and clearing deviations. Because so many pharmaceutical businesses have not properly analyzed these costs, they are unaware of the potential savings. This subject is pursued further in Chapter 16.

Some management aspects central to a successful Six Sigma approach are the following:

1. Senior managers must not only be the leaders of Six Sigma, but they also must be seen as the leaders. Visibility and credibility are essential. "Follow me" is an approach that works. "I'm right behind you" is one that seldom does.
2. Every manager must be held accountable to achieving Six Sigma levels of variability, with particular reference to process owners. Process owners must be accountable for the entire process for any given product or service, not just a department or similar element.
3. Management must provide education and training for everyone in the organization. The level of education and training will vary depending on the role each individual plays, but everyone needs a basic understanding so they are able to participate.
4. Resources and, in particular, trained full-time improvement project leaders must be allocated to business units to drive the improvement strategies.
5. Information systems need to be developed to track progress. This is not an additional system; it is a revamp of the core information system. For key items, monthly variance reporting should be replaced with control charts so managers can differentiate between random variation and signals in the data.
6. In the medium to long term, the discovery, development, and design elements are the most important areas to address.

Any organizational psychologist will explain that most people need structure in their working lives. Many of the companies that stumbled with their initial attempts to introduce Deming's approach, Kaizen, total quality management (TQM), etc. discovered that part of the problem was that they had failed to put into place structures that provided guidance and cultural guideposts as well as some of the "how to." While it is certainly possible to create a snarl of counter-productive bureaucracy, Six Sigma demonstrated how these structures could be helpful in companies such as Motorola and GE.

A widely misunderstood part of Six Sigma structures concerns leaders and black belts. Black belts are widely thought to be what Harry calls "tool masters," or people skilled in the use of sophisticated statistical methods. Harry writes that leaders and black belts should be idea mongers rather than tool masters.[7] It is in this spirit that this book is written. The ideas or concepts should drive the structures and tools, rather than the reverse.

Some of the salient structural elements introduced as part of the Motorola Six Sigma approach include:

1. Removal of options to participating.
2. Allocation of substantial resources.
3. A project-by-project approach similar to the one about which Juran wrote so passionately.
4. Accountability and a demand for results measured in terms of both reduced variation and financial outcomes.

Options. One of the very clear features of Motorola's approach was the removal of options to effectively participating in their Six Sigma initiative. This was part of their "shotgun" approach. Everyone was expected to participate. During his early seminars, Harry was fond of telling stories about managers and executives who failed to participate fully, and of the "significant emotional event" with a vice-president that invariably followed.[3] A favorite story comes from the CEO of an Australian company introducing a quality initiative in the 1980s. He would tell the old story about the production of his favorite breakfast, bacon and eggs. The two animals involved were a chicken and a pig. The chicken was *involved* in the production of his breakfast, but the pig was *committed*. He would then explain that his approach would work only for pigs, and that chicken managers were not welcome in his business. It is a useful analogy. Six Sigma only works for pigs.

Resources. Another element that tended to be missing from many other improvement approaches was the allocation of significant resources. The green and black belts may seem odd to some, but even those not enamored of the terminology would mostly applaud the allocation of resources to help managers conduct improvement projects. A course or seminar on the subject followed by an exhortation to sally forth and "do Six Sigma" is not sufficient training and guidance for most managers. In his book, *Out of the Crisis,* Deming lays out a structure that could easily be mistaken for the approach developed at Motorola.[8] These similar approaches have stood the test of time. Unfortunately, in the words of the Bard, they seem to be more honored in the breach than in the observance.

Project-by-Project. Those familiar with Juran's work will recall that he was adamant that no other approach would work. (Initially, Juran and Deming seemed to differ on this subject to some extent. It is a position on which both softened a little later in life.) I see no need for the project-by-project approach and the more general cultural and shotgun approaches to be mutually exclusive. May not both be used? Must this be an either-or argument or might it be a suitable place for an "and" approach? Motorola certainly opted to pursue both approaches simultaneously.

Accountability. Another element of the Motorola method was their approach to accountability. The company made it clear that it believed that conquering variation would lead to improved financial and customer service outcomes. To this end, the accountability model was altered so that a prime measure of a manager's success was his ability to improve products, services, and processes in Six Sigma terms; that is, to reduce variation and to show demonstrable improvements in business results. At his early seminars, Harry would produce a slide showing a distribution in the crosshairs of a telescopic riflescope. "You've got to shoot the tails of *all* your distributions," he would say, followed by, "...and managers must be held accountable to reduce variation to Six Sigma levels." Problems and costs in manufacturing that could be traced back to design or facility construction were transfer-priced back to the design and construction departments. This fundamentally changed the perspective of many managers. So too did the demand that financial people be involved in the planning and execution of major projects so that high value improvements were targeted and improvements could be better quantified.

CONCLUSION

The concept called Six Sigma was born in Motorola in the 1980s, but it has a long and proud tradition that reaches back to the 1920s, as will be noted in Chapter 4. It brings some interesting new ways to explain why reducing variation is a business imperative. Also, it provides structure to assist with the "how to" and a sound array of tools and process improvement methods. Sadly, in some consulting firms and businesses the Six Sigma concept seems to have grown from its original roots into a tangled thicket of paperwork, "administrivia," and management ritual. Structure is not a bad thing, but too much structure and bureaucracy will choke even the best of ideas. As Harry noted, ideas and concepts are more important than tools or structures. If the tools and structures become more important than the core ideas, the ideas can be smothered.

Six Sigma is a tried, tested, and proven approach. Nonetheless, some have modified the original approach to such an extent that it is barely recognizable. Elements that are difficult to teach or to sell are omitted. Branches cut from exotic species are grafted onto the rootstock. Sometimes, this robs the organism of its vitality. In many cases, the approach is burdened with so much bureaucracy that it becomes a ticket punching exercise. No good idea is proof to such tampering. This was noted in the past with the work of Deming, Ohno, and Juran, for example, and with approaches such as re-engineering and supplier partnering. It seems criminal when an approach with the enormous potential of Six Sigma becomes hamstrung with bureaucratic nonsense, but it can happen unless leaders are vigilant.

However, if readers are aware of the original concepts they can defend themselves against misinformation. The clue is in the name, Six Sigma. It is, or should be, a statistical approach. Its foundation is reducing variation and managing systems.

ONE POINT LEARNING

1. Six Sigma is a statistical concept. Its application requires an ability to understand and reduce variation.
2. The empirical evidence is overpowering. Low variation decreases costs and increases revenues.
3. Six Sigma addresses the entire process, across departmental boundaries.
4. Six Sigma contains both an organizational culture (shotgun) element and an improvement project (sniper rifle) element. Both should be used.
5. Organizational structures should promote rather than inhibit the Six Sigma approach. This includes reporting and information systems, and the formal allocation of resources to the organizational structure to conduct improvement projects.
6. Discovery, development, design, and construction areas have the greatest impact on performance.
7. Six Sigma only works for pigs.

REFERENCES

1. M. Harry and R. Schroeder, *Six Sigma*, Currency Doubleday, New York, 2000.
2. D.J. Wheeler and D.S. Chambers, *Understanding Statistical Process Control*, Addison-Wesley, Boston, MA, 1990.
3. *Six Sigma Seminar*, sponsored by Motorola University, M. Harry et al., Sydney, 1991.
4. G.E.P. Box, J.S. Hunter and W.G. Hunter, *Statistics for Experimenters: Design, Innovation and Discovery*, 2nd ed., Wiley, New York, 2005.
5. *A Prophet Unheard*, Videotape, BBC Enterprises, London, 1992.
6. W. Hopp and M. Spearman, *Factory Physics*, McGraw-Hill, Boston, MA, 1996.
7. M. Harry, *Six Sigma Knowledge Design*, Palladyne, Phoenix, AZ, 2001.
8. W.E. Deming, *Out of the Crisis*, MIT Press, Cambridge, MA, 1988.

3 Evolution

Some Traditional Concepts of Variation

Our understanding of variation has evolved over hundreds of years. Understanding this evolutionary process helps us to understand from where our existing approaches came. It also helps us to understand the inevitability of the next stage in the journey.

One of the first laws of nature is that no two things are identical. No two chromatographic columns are exactly alike. No two batches produced in a chemical or biological process are quite the same. Every day people in the pharmaceutical industry struggle to understand and explain variation and deviations. This struggle can be traced from the origins of manufacturing to the present. An enormous amount of time, money, and energy continues to be invested in this effort. In order to understand from where our current practices spring, as well as the superior approaches that have evolved in recent times, it is instructive to take a historical perspective.

IN THE BEGINNING...

In the beginning, craftsmen made all products. Usually, the craftsman made the entire product. There were no subcontractors or vendors. The baker ground the wheat, collected, processed and added all the ingredients, built his fire with wood his children collected, and produced unique bread. The armorer even made his own rivets. The blacksmith made his own nails. Every product was unique. Parts were not interchangeable because every part was individually crafted to match other components, so that when assembled, they made a workable whole. Archers made arrows that were designed and fabricated to match each longbow, which was itself, quite unique. A blacksmith might make two winches for a castle's drawbridges. However, if one winch component broke, a like part could not be scavenged from the other winch because each component was hand crafted to match the intermeshing components. In such an event, a new part had to be made to order.

THE ADVENT OF MASS PRODUCTION

For a long time, early industrial thinkers dreamed of the interchangeability of parts, and what a dream it was. If parts could be made so they were interchangeable, a whole new era of manufacturing would follow. It would allow components to be mass-produced and then assembled as a separate function, by different people, sometimes on separate continents. Manufacturing costs would be reduced dramatically and living standards would soar to unheard of heights.

History has proven that these early thinkers dreamed soundly. Any teenager who has tinkered with cars knows that he can scavenge parts from a like vehicle to repair his own. The idea that components could be salvaged from a broken product and used to repair another was not apparent to people at the time of the industrial revolution. Modern manufacturers find it difficult to understand how this would not be patently obvious. However, many things that are obvious with the benefit of hindsight were anything but clear at the time. Before people knew that interchangeability of parts was possible, how could they have envisaged the teenager scavenging parts from a broken product to repair his own?

The evidence that manufacturers overcame these obstacles surrounds us. Mass production would not be possible without interchangeability of parts. The benefits brought to humanity by this lack of variation defy measurement. Without mass production, there would be no cars, refrigerators, computers, packaged food, or household appliances. There would be no mass transit systems, reticulated water, or sewage systems. The medicines humans need and the means of administering them would not be freely available.

The same is true of batch and continuous processes, which are commonly found in the pharmaceutical industry. Most drugs and vaccines are made in batches, through a series of processing events. This form of manufacture is different from that found in plants that make components for cars, computers, or cell phones, but the same principles of mass production apply. Raw material is sourced, measured, and mixed. A molecule is created, sometimes by a chemical reaction or sometimes it is grown biologically. Several more steps to modify, isolate, and purify the molecule are common, as are other events to fold, clip, or otherwise transform the molecule into a usable form. Finally, it is formulated and packaged into doses. Usually, this happens batch by batch. The pharmaceutical manufacturer tries to closely replicate each batch, time after time. This is mass production, except it is the batch that needs to show interchangeability with other batches rather than the component-to-component interchangeability found in plants manufacturing medical devices.

In December 1979, the World Health Organization declared smallpox to be eradicated. The virus survives in two secure labs only. It is difficult to believe that this triumph could have been achieved without cheap, mass-produced vaccines.

In laboratories, the interchangeability of interest is between analysts and instruments. If a blind control sample gives very similar results regardless of which analyst conducts the test or which instrument is used, genuine interchangeability has been achieved.

Everywhere one looks, the same principles apply. The industrial revolution ushered in mass production based on interchangeability. In terms of advancing the material welfare of humans and of lengthening our lifespan, it was a staggeringly huge achievement.

Some of the early attempts to achieve interchangeability were centered on firearms manufacture. Manufacturers tried to introduce mass production, but initially they struggled. When the separate components were brought together for assembly, often the fit was poor. Sometimes, the fit was so poor that parts were rejected as rework or scrap.

Manufacturers began to realize that exact repeatability was not possible. Soon after, they recognized that it was not necessary. What was needed was for the variation between parts to be low enough such that the assembled product was functional. The first breakthroughs were the invention of "go" and "no-go" gauges. Imagine the boring out of barrels for firearms. The hole bored must be big enough to allow the bullet to be inserted and rammed home, but not so big that on firing the gases escape around the projectile. In this case, a bar of metal with one end machined to a slightly smaller diameter becomes the gauge. The small end must fit into the barrel, but the larger end must not. Before long, manufacturers had converted the go, no-go gauge into the concept of tolerances or specifications.

Now manufacturers had a way of determining in advance whether the parts made could be assembled successfully into a workable product. This created the first definition of quality:

All parts must remain within the go, no-go specifications.

Of course, Six Sigma calls for far less variation than this. While it is now easy to see the shortcomings in this definition, at the time it was thought to be satisfactory. Bear in mind that manufacturing processes were far less complex than they tend to be today, and in that circumstance variation had less impact. This is one reason why an approach that may have worked reasonably well at the time no longer serves our needs.[1] Specifications, in diagrammatic form, are illustrated in Figure 3.1.

This method of determining the quality of a product is as simple as it is widespread. All products that fall within the specifications are okay. All those outside specifications are rejects. This is a black and white world. There is no grey. It is a

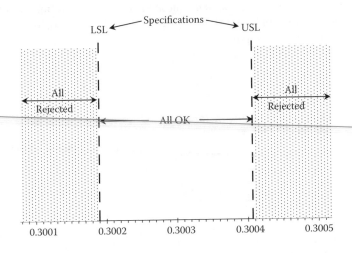

FIGURE 3.1 Go and no-go specifications for a firearm bore.

simple world, but unfortunately, as will be shown in later chapters, the pharmaceutical industry does not live in that world.

Unfortunately, ever since the first specifications were written, products that failed to meet them have been made. Such products cause much economic loss and pharmaceutical companies spend a great deal of time and money raising and clearing deviations as well as attempting to prevent future recurrences. In many instances, despite years of investigations and corrective action, the rate of deviations remains reasonably constant or increases. This is not because of regulatory "creep," but is attitudinal in origin, as will be shown later.

ILLUSTRATING VARIATION

THE FREQUENCY DISTRIBUTION

It is common practice to measure products and processes. Unfortunately, some pharmaceutical operations are data-rich, but analysis-poor. To provide an example, imagine the active ingredient in certain tablets is being measured immediately after the tablets are compressed into shape. Samples are tested and the data in Table 3.1 are assembled.

The tabulated format of Table 3.1 is the most common way of assembling data, but this approach, while convenient for storing large quantities of data, provides no analysis. Perhaps the simplest analytical method for data such as these is to create a frequency distribution as seen in Figure 3.2.

The distribution in Figure 3.2 shows the frequency with which each measurement was found in the sample. Three characteristics in which one should always be interested are the range, average, and shape. These are marked on the chart in Figure 3.2. The distributions in Figure 3.3, Figure 3.4, and Figure 3.5 illustrate how distributions may vary in range, average, and shape.

Suppose some of the active ingredient data from Figure 3.2 and Figure 3.3 were outside of specifications. The obvious question is, "Why?" Are the out-of-specification tablets failing in range, location, or shape? Figure 3.6, Figure 3.7, and Figure 3.8 illustrate some possibilities.

TABLE 3.1
Active Ingredient in Tablets

754	757	756	752	756	757	754	755	753
756	761	754	757	755	756	759	755	755
755	758	754	758	756	754	753	757	754
752	756	755	754	754	752	758	753	752
759	752	750	761	761	754	756	754	759
756	753	756	757	759	758	754	759	751
757	756	757	755	754	762	754	756	755
760	756	760	753	755	757	757	755	754
757	755	755	754	753	760	758	758	759
753	755	756	756	752	756	757	759	754

		Range = 11	Frequency
750	X	Average = 755.6	1
751	X	Shape = Bell curve	1
752	X X X X X X		6
753	X X X X X X X		7
754	X X X X X X X X X X X X X X X X X		17
755	X X X X X X X X X X X X X X		14
756	X X X X X X X X X X X X X X X		15
757	X X X X X X X X X X X		11
758	X X X X X X		6
759	X X X X X X X X		8
760	X X X		3
761	X		1
			Total 90

FIGURE 3.2 Frequency distribution for active ingredient in tablets.

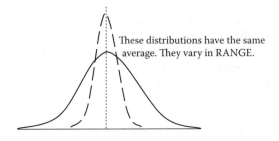

These distributions have the same average. They vary in RANGE.

FIGURE 3.3 Distributions varying in range.

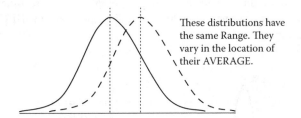

These distributions have the same Range. They vary in the location of their AVERAGE.

FIGURE 3.4 Distributions varying in the location of their average.

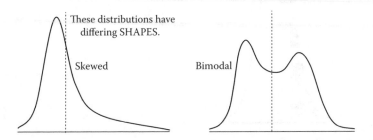

These distributions have differing SHAPES.

Skewed

Bimodal

FIGURE 3.5 Distributions varying in shape.

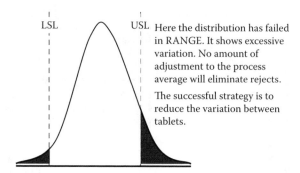

FIGURE 3.6 Failure caused by excessive range.

Case Study — Chromatography Yields

In this example, chromatography was used to capture a therapeutic protein from a fermentation broth. The initial yield data was unstable but indicated that different people repacked and operated the chromatography columns quite differently. The distribution for several runs was bimodal. The first step was to conduct training and to enforce standard procedures for repacking and operating the columns to create uniformity between all operators and technical people.

As can be seen in Figure 3.9, the result of this work was not only reduced variation, but also a higher average yield. Subsequent work to reduce variation even further increased the yields yet again. Both yield and capacity increased once the process was uniform, more repeatable. Unit costs fell. Just as importantly, the deviation rate was reduced and the technical people had more time to work on improving the process rather than having their time consumed by raising and clearing deviations.

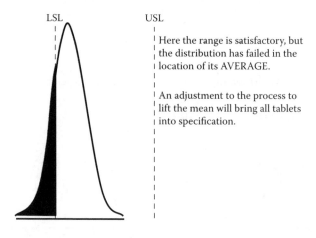

FIGURE 3.7 Failure caused by wrong location of the average.

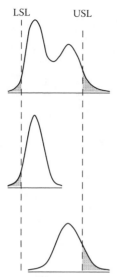

This distribution is bimodal in SHAPE. It is not a single process. Rather it is two processes whose data have been mixed together. In this case the shape was caused by the morning and afternoon shifts operating differently.

This is the afternoon shift. Its AVERAGE is in the wrong location.

This is the morning shift. It shows excessive RANGE. It cannot be centered in such a way that no rejects will be made.

FIGURE 3.8 Shape can reveal causes of excessive variation.

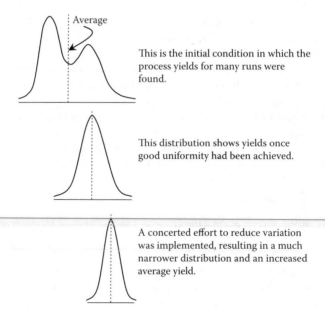

This is the initial condition in which the process yields for many runs were found.

This distribution shows yields once good uniformity had been achieved.

A concerted effort to reduce variation was implemented, resulting in a much narrower distribution and an increased average yield.

FIGURE 3.9 An example of process improvement — chromatography yield.

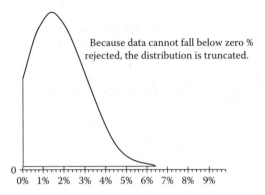

FIGURE 3.10 Truncation caused by a natural limit.

TRUNCATED DISTRIBUTIONS

Distributions that are truncated, or that appear to be "chopped off" are not uncommon. There are several potential reasons for this characteristic, and three are discussed here:

1. There is a natural limit past which the data cannot fall. For instance, it is not possible to have fewer than 0% defectives. This limit may make a distribution appear truncated, as noted in Figure 3.10. This is also true for parameters where the analytical data fall below the method limit of quantitation (LOQ) or reporting limit.
2. A situation exists where products falling outside specifications are found and rejected during inspection. The removed data can cause the distribution to appear truncated, as noted in Figure 3.11.
3. Data are corrupted such that out of specification results are not recorded, or are adjusted so that they appear to meet specification. Thankfully, the authors have yet to see such an occurrence in the pharmaceutical industry. Figure 3.12 shows an example from the food industry.

THE NORMAL DISTRIBUTION

The normal distribution, or normal curve, is an exact statistical model that possesses certain rigid characteristics. It is found only in textbooks. An exactly normal curve has never been found in our world. The best that can be hoped for is a distribution of data that is close to normal. Dr. George Box's quote, "All models are wrong; some models are useful," is worth remembering. The normal curve is at least of the useful variety. Later chapters will explain the limitations to its use. Its characteristics include the following:

1. It has a symmetrical bell shape about its average.
2. It can be completely described mathematically.

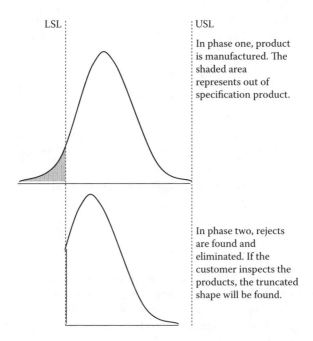

FIGURE 3.11 Truncation caused by eliminating rejects.

3. Its range is always about six standard deviations, with only 3 data in 1000 falling further than three standard deviations from the average.
4. The proportion of data that fall under any particular section of the curve can be predicted exactly. Figure 3.13 shows the percentage of data that fall between one-sigma graduations, rounded off to the nearest whole number.

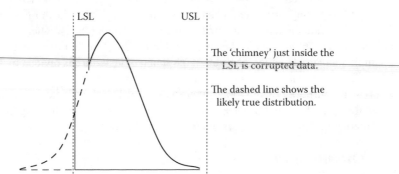

FIGURE 3.12 Truncation caused by corrupted data.

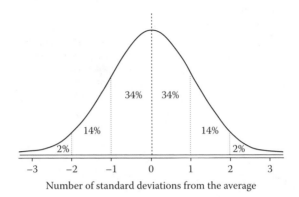

Number of standard deviations from the average

FIGURE 3.13 The normal curve.

The full name of standard deviation is *root mean square deviation*. The key word is *deviation* because sigma is a measure of variation, the tendency of the data to disperse around its average, or a measure of the tendency of the data to deviate from its average. There are several methods to calculate sigma. To illustrate one, suppose one wished to calculate sigma for the data 2, 3, 4, 5, and 6. The formula used is:

$$\sigma = \sqrt{\frac{\sum (x_1 - \bar{x})^2 + (x_2 - \bar{x})^2 + \quad (x_n - \bar{x})^2}{n}}$$

where x is the individual result, $\bar{x}$ is the mean or average of the data set (4 for this example), and n is the number of data in the sample.

The standard deviation or σ of the data 2, 3, 4, 5, and 6 is calculated as 1.414. A similar statistic is the sample standard deviation, usually denoted as "s" or as $\sigma_{(n-1)}$. This statistic is often used when dealing with small sample sizes to help reduce the impact of sampling error. The calculations are very similar to those seen above, except when calculating a mean, one divides by a number one smaller than the n or number of data in the sample. That is, one divides by $n - 1$. In this case, to calculate a sample standard deviation, the sum of the squared deviations is divided by 4 instead of 5. This would yield a sample standard deviation, s, or $\sigma_{(n-1)}$, of 1.58.

For the purposes of this book, there is no need to delve into any more detail about the normal curve. For those readers who are interested in pursuing the subject, the Internet has many references and good texts abound.

TIME ORDERED DISTRIBUTIONS

Distribution theory is very useful. However, it buries the information contained in the time-ordered sequence of the data. One (imperfect) way of observing a process over time is to use a series of time-ordered distributions. When this is done in

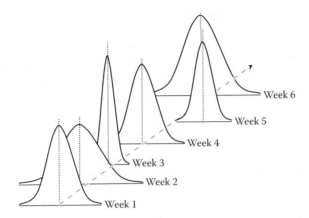

FIGURE 3.14 Unpredictable time ordered distributions.

pharmaceutical factories and laboratories, the most common outcome looks similar to the diagrams in Figure 3.14.[2,3]

The distributions in Figure 3.14 illustrate a most common problem. They are unpredictable, perhaps chaotic. If a plant's processes look similar to Figure 3.14 how does anyone effectively forecast and schedule production? How are meaningful budgets and business plans prepared? If laboratory controls resemble this chaotic pattern, how confident can one be about the results from the production samples? For as long as the process behaves so recklessly, production scheduling and providing guarantees of supply can be very problematic.

In addition to these issues, when the data are chaotic it is either difficult or impossible to effectively determine cause and effect relationships. For example, assume a change had been made at the end of week 2. In week 3, a much-improved outcome is noted. Variation has been reduced. However, for as long as the process remains so uncontrolled and erratic, no one can know what caused this most desirable result with any surety.

Matters would become even more complicated if specifications were added to the diagram in Figure 3.14. Some days there are no or few rejects. On other days, some results fail the upper specification, and some fail the lower specification, or both. The Six Sigma concept demonstrates that relying on knowing whether outcomes meet specifications or standards is a very poor approach to understanding and controlling a process. Later chapters will introduce superior approaches to understanding data.

For a long time, knowledge of variation and its effects in the pharmaceutical world stalled at this level of understanding. The apparently reckless behavior of production data, product distribution data, and lab controls baffled chemists, biologists, engineers, and managers.

It is now clear that only when managers and technical people understand the process through the application of statistical tools that the potential of the process will be revealed and realized, by reducing and controlling variation.

ONE POINT LEARNING

1. The go, no-go approach is perfectly satisfactory for release specifications, but totally inadequate for understanding or controlling a process.
2. Distribution theory is useful, but superior methods that show how the process performs over time are required.

REFERENCES

1. W.E. Deming, *Out of the Crisis*, MIT Press, Cambridge, MA, 1988.
2. E.L. Grant and R.S. Leavenworth, *Statistical Quality Control*, McGraw Hill, New York, 1980.
3. W. Scherkenbach, *The Deming Route to Quality and Productivity*, CEEPress, Washington, D.C., 1988.

4 Revolution

The New Understanding of Variation

Dr. Walter Shewhart revolutionized our understanding of variation. His control charts were a breakthrough, but his ability to perceive the universe through new lenses and to understand that which hitherto had been concealed from science was profound.

Shewhart studied variation during his work at Bell Telephone Laboratories in the 1920s. His studies included variation observed in nature, as well as that found in manufacturing environments. He observed that in nature, and in certain controlled experiments, he could predict the limits of variation in samples taken from a given population. However, in manufacturing he noted that the samples he studied rarely followed a predictable pattern. Many were wildly erratic and unpredictable. In the pharmaceutical industry, this erratic behavior leads to failed batches, repeated tests, raised costs, damaged quality, and deviations.

Shewhart observed that some manufacturing processes were quite predictable, and he sought to understand why some processes were predictable or controlled, and why others were not. Early in these studies, he formulated a theory that may be summarized as follows:

All processes display variation.

Some display controlled (or stable) variation.

Some display uncontrolled (or unstable) variation.[1,2]

Shewhart knew that in any process many variables exist. People, machines, materials, methods, measurement systems, and the environment all combine in an interactive way to produce outcomes that vary. He discovered that it was possible to differentiate between stable and unstable variation, as outlined in the following paragraphs.

Stable variation. Some variation was unavoidable because the variables interacted in different combinations from time to time. When this variation was uniform and predictable over time, it was due to the randomness brought about by the chance interaction of the variables. Shewhart called this variation due to chance causes *controlled variation*. Such a result, if plotted as a time series of frequency distributions, would appear as seen in Figure 4.1, and the process is said to be in a controlled state, in statistical control, or stable.[3,4] Because the word *control* has a specific meaning in management jargon, generally the term *stable* is preferred to avoid confusion.

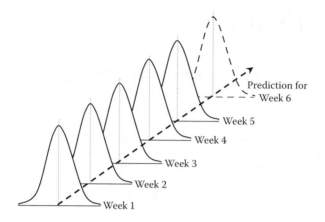

FIGURE 4.1 A stable process.

Unstable variation. In addition to the randomness caused by chance interaction of variables, there are other sources of variation that may impact on the process but which follow no predictable pattern. For convenience, these sources can be broken into two categories:

1. **"Assignable" or "special" causes of variation.** These are the causes of variation that usually can be traced to a specific operator, shift, instrument, or batch of raw material. This variation is over and above the background noise of randomness. "Special" does not mean undesirable, it means different. If one analyst is operating differently from the remainder, her results are likely to look "special" when compared to the other analysts.
2. **Regular changes to the process.** This issue is common in batch processes in the pharmaceutical industry. It occurs, for instance, when changes to the process parameters for the next batch are made based on recent results.

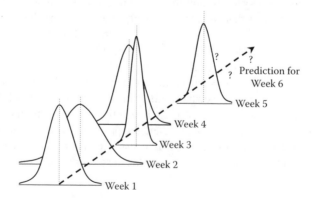

FIGURE 4.2 An unstable process.

This can be done by both technical people and operators. Significant over-control or unnecessary adjustment is common in the pharmaceutical industry and will be discussed in detail in Chapter 6, Chapter 13, and Chapter 14. In strict terms, these causes of variation are also "assignable," but they are treated separately here for ease of explanation.

Regardless of whether the process is impacted by special causes or regular process changes, the process will become unstable when these occur. Figure 4.2 illustrates this condition.[3,4]

IS THIS UNDERSTANDING IMPORTANT?

Why should anyone be concerned about whether the process is stable or unstable? Is there a price for ignorance? At a fundamental level, there are two reasons why everyone ought to be concerned if their processes are unstable. These issues are centered on the notions of predictability and establishing cause and effect relationships.

Predictability. If a process is unstable, by definition it is not predictable. Without adequate prediction, how does any business build business plans, budgets, forecasts, or production schedules? If it is found that production schedules are constantly revised, or that business planning and forecasting is inordinately difficult, chances are the processes are unstable. It is self-evident that the better the process can be predicted, the easier will be the forecasting, planning, scheduling, and daily or weekly management. Moreover, everything done under the banner of management requires some prediction. To the extent that managers and technical people cannot predict, they cannot manage; they react. Management ought to be the opposite of reaction, at least for the most part. Do managers and technical people control the process or does it control them? If it is found that reacting to unplanned issues is common, it is likely the relevant processes are unstable.

Cause-and-Effect Relationships. It is a fundamental principle of statistics that cause-and-effect relationships cannot be effectively demonstrated while the process is unstable. Consider the situation in Figure 4.2. If the process owner had made a change to the process at the end of week 2, how sure could he be that the improvement noted in week 3 was a function of his actions? As long as the process remains unstable, he will never know with any surety. If people struggle to grasp cause-and-effect relationships, improvement is likely to be slow, where it exists at all. Our experience here is uniform. If instability exists, even the best technically trained people will struggle to understand and improve it. Several examples of this truth will be offered in later chapters.

STABILIZE FIRST!

One of the most important messages in this book is that one should not attempt to improve an unstable process in engineering, biological, chemical, production volume and quality, or financial terms. Some of the most talented people in the pharmaceutical industry have shattered themselves against the rocks of instability. The process is unpredictable…the causal relationships are not clear…the results from plant trials

are ambiguous…deviations are common. No amount of hard work or technical talent
will overcome these problems because they are seldom of a chemical, biological,
or engineering nature. Nearly always they are operational. They are related to how
the process is operated, day by day, shift by shift, moment by moment.

If the process is unstable when first examined, the first priority ought to be to
stabilize it; to make it repeatable and predictable. Only then will causal relationships
become clear. Only then will trial data be unambiguous. Only then will the managers
and technical people be in a position where they control the process rather than
having it control them.

…THEN IMPROVE THE PROCESS

Once the process is stable, it will be much simpler to establish causal relationships
and to improve plant performance. The work done in this phase is more likely to
be technical, but now the technical people will have become empowered. They
will be able to determine which changes work and which do not. Planning,
scheduling, and forecasting become simpler. Finally, they control the work. It does
not control them.

THE FIRST PRINCIPLE

"Stabilize first" *is* the first principle. The importance of this approach cannot be
overstated. The two-phased approach to process improvement is central to the ideas
first developed by Shewhart. First the process is stabilized. This means making it
as exactly repeatable, shift by shift and batch by batch, as one can imagine. *Then,
and only then, should one embark on the improvement process.* The only proviso is
that if the wrong product is being made or inappropriate technology is in use, these
aspects need to be addressed immediately. There is no point in making a uniform
product with a stable process if it is not what the market wants, or if the technology
is obsolete and needs to be replaced.

Instability is the fog that shrouds the capabilities of the process. By lifting this
fog, one opens the door to better forecasting, planning, and scheduling; one can
better expose and enumerate causal relationships, and one can create new possibilities for the future.

From this new understanding of stability comes the second definition of quality:

All characteristics must remain stable and within the go, no-go specifications.

DEMING POLISHES THE DIAMOND

It is a popular part of our folklore that one must sift through many tons of gravel
to find a single diamond. Shewhart's theories and control charts were diamonds in
the gravel pit of management practices and fads. Unfortunately, these diamonds were
uncut and unpolished in the sense that most biologists, chemists, and engineers were
about as excited by Shewhart's work as an untrained eye would be about an uncut
diamond. Most people missed their potential.

Another American statistician, Dr. W. Edwards Deming, had a trained eye and saw the huge potential Shewhart's methods offered. The two men were good friends who worked closely together. Deming arranged lectures and seminars for Shewhart. He edited and helped to arrange the publishing of Shewhart's writings. Over the years, Deming was to develop an entire management philosophy partially based on the new statistical methods and incorporating them. The standard text for Deming's work is *Out of the Crisis*.[5]

DEMING'S FIRST OPPORTUNITY

The use of Shewhart's theory and methods had not become widespread by the time the U.S. entered World War II. Deming and others worked hard to have Shewhart's methods introduced into industry to boost wartime productivity. Lectures and courses were given across the U.S. and by the end of the war many thousands had received training in the new statistical methods.

Sadly, when the war ended, so did the large-scale study of Shewhart's new statistical methods. The U.S. was an industrial giant riding the crest of an economic wave. The world was hungry for products from the U.S. Anything the U.S. could make, she could sell. The quality approach gave way to a quantity approach, partly because Deming's and Shewhart's teachings had been received primarily by technical people. Senior managers did not understand Deming's message. They did understand that the world was a seller's market, that they were making healthy profits, and that there was more pressure to improve quantity than quality. In this environment, the scene was set for superstitious knowledge to evolve...and it did.

DEMING'S SECOND OPPORTUNITY

In the late 1940s, Deming went to Japan to help the Japanese with some statistical work. He met and impressed many Japanese businessmen, government officials, and statisticians. Deming was invited to Japan in 1950 to give a series of lectures and seminars on the new approach to variation, quality, productivity, and market research. The singular hope of the Japanese was that Western experts such as Deming could help them to return to pre-war levels of productivity and prosperity.

Deming, however, had other ideas. He told the Japanese that if they adopted his approach to management and quality, then within a few years Western businesses would be pleading with their governments for protection. Readers may judge for themselves the success or otherwise of Deming's approach in Japan. Nevertheless, it is noteworthy that to this day one of the most prestigious awards one may receive in Japan for quality and productivity is the Deming Medal.

THE DEMING APPROACH

The heart of the message Deming took to Japan in 1950 is as simple as it is profound. The simple flow diagram that Deming wrote on the blackboard in 1950 has since been shown thousands of times all over the world. It is reproduced in Figure 4.3.[5]

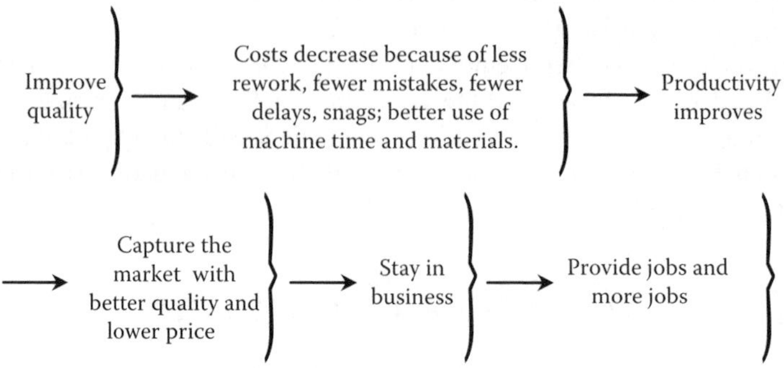

FIGURE 4.3 Deming's chain reaction.

To fully appreciate the meaning of Deming's flow diagram, an understanding of his definition of quality is required:

Good quality does not necessarily mean high quality. It means a predictable degree of uniformity and dependability, at low cost, suited to the market.[6]

A key phrase in this definition is *a predictable degree of uniformity*. This can be expressed as *a predictable degree of variation*. Deming was saying that good quality was not necessarily about excellence. Rather, it was about producing product that was uniform and predictable, day to day and batch to batch, as well as being satisfactory from a customer's perspective. This type of quality is going to spring only from stable processes. Shewhart's control charts are significant in assisting the understanding, stabilizing, and controlling of variation. Deming then took the concept one step further with the following concept:

It is good management to continually reduce the variation of any quality characteristic, whether this characteristic be in a state of control or not, and even when few or no defectives are being produced.[1]

The traditional view of variation is quite different, and it persists to a large degree in highly regulated industries such as the pharmaceutical industry. Because the pressures to meet compliance standards and specifications are so intense, the go, no-go mental model that sorts outcomes into good and bad, meets or does not meet specifications, is ingrained into the psyche of most people in the pharmaceutical industry. These compliance issues will not go away, but they need not blind us to another way of understanding and reducing variation that offers the promise of far superior compliance, fewer deviations, and improved productivity.

Later chapters will pursue further the issue of being content with merely meeting standards, as this problem is fundamental to the pharmaceutical industry.

LIMITS TO KNOWLEDGE

One must bear in mind that words such as *exactly*, *pure*, *reliable*, and *precise* have no meaning until the methods to be used for sampling and measurement have been decided upon.[5] If one changes the method of inspection, sampling, or testing, new numbers are created.

The laboratory never tests the process or the product. It tests the sample. Sampling error can be significant. How well is it understood? How and by what method? What is the test error? How well is the analyst-to-analyst and instrument-to-instrument error understood? Are blind controls used to measure the total analytical error, including aspects such as changes to reagent batches and different analysts on different shifts using differing instruments?

There are limits to knowledge of any process, but if the sampling and testing processes are stable, the potential error will be constant over time and can be taken into account.

Shewhart's approach is based on statistical analyses, but any statistical method can be confounded if the sampling and testing processes are unstable. As will be explained in more detail later in this book, it is indeed rare to find a laboratory in the pharmaceutical industry where the initial analyses indicate that the laboratory controls are stable. What does that say about the confidence that may be claimed for production samples? How does this impact compliance capability? Stability does not just happen. It is a consequence of a deliberate aim and determined effort.

ONE POINT LEARNING

1. Stabilize first *is* the first principle.
2. Six Sigma is not necessarily about excellence or high quality. It is always focused on reducing variation to a minimum.
3. Understanding the variation in the sampling and test methods is vital information.

REFERENCES

1. W.A. Shewhart, *Statistical Method from the Viewpoint of Quality Control*, The Graduate School of Agriculture, Washington, D.C., 1939.
2. W.A. Shewhart, *Economic Control of Quality of Manufactured Product*, Van Nostrand, New York, 1931.
3. E.L. Grant and R.S. Leavenworth, *Statistical Quality Control*, McGraw-Hill, New York, 1980.
4. W. Scherkenbach, *The Deming Route to Quality and Productivity*, CEEPress, Washington, D.C., 1988.
5. W.E. Deming, *Out of the Crisis*, MIT Press, Cambridge, MA, 1988.
6. W.E. Deming, 4-Day Seminar, Sydney, 1986.

5 Paradox

The Parable of the Red Beads

It is paradoxical that seemingly chaotic data can in fact be stable, predictable. Deming's famous Parable of the Red Beads is devastatingly simple in its explanation of this paradox.

An experiment using statistical sampling beads provides an excellent illustration of Shewhart's approach. This experiment was made famous by Deming.[1,2] The equipment the authors used is as follows:

1. 3000 white beads and 600 red beads.
2. Two containers to hold and mix the beads. Each container is less than half full when holding all 3600 beads.
3. A sampling paddle with 32 holes. A sample taken by dipping the paddle into the container takes a sample of 32 from a population of 3600.

A scenario is set where a manufacturing business is being established. A staff of six willing workers, an inspector, and a recorder are recruited from the audience by the manager. The workers are briefed to the effect that their job is to produce white beads, and that red beads are defectives. The inspector's job is to count the number of red beads in each worker's production and announce this count to workers and audience alike. The recorder's job is to record the results on a tally sheet.

A uniform method of making a batch (taking a sample of beads) is taught to the workers, and this method is strictly enforced to eliminate variation in work methods. The beads are mixed between workers by twice pouring them from one container to another. This method is not perfect but is good enough for the purpose of the experiment.

The workers are briefed to the effect that their company is a high-quality company. Shoddy work will not be tolerated. They are told that those who perform well will be rewarded. A financial incentive is created, and the workers are informed that poor performers will be removed. Therefore, their jobs depend on their own performance (the well-known principle, Perform Or Get Out). A limit of three red beads per worker per day is set.

The exercise is conducted. The workers take turns producing batches and their results are recorded. The results of one such example can be seen in Table 5.1. At the end of the exercise, the three workers who produced the highest number of red beads are fired. The cash prizes offered as the incentive are presented to the workers who produced the least number of red beads, usually among the applause of the audience, despite the fact that most are well aware that the prizes have been awarded on a lottery basis.

TABLE 5.1
Results of Red Beads Experiment

Worker	Day 1	Day 2	Day 3	Day 4	Total
Tim	5	7	11	6	29
Shane	2	6	5	6	19
Jake	6	7	2	5	21
James	1	3	5	3	12
Brian	9	8	6	9	32
Annabelle	5	5	8	2	20
TOTAL	28	36	38	31	133

Manager: Darcy   Inspector: Danielle   Recorder: Nathan

After the financial rewards have been distributed, the three best performing workers are retained and the others are dismissed. The three who remain will work double shifts for four more days. For many, it makes a lot of sense to remove the dead wood and make best use of the top performers. The second phase results for the group used in this example can be seen in Table 5.2.

The audience then formed into small syndicates to consider and respond to these questions:

1. Was the experiment an example of a stable system?
2. Did the inspector, recorder, or manager help at all to improve matters?
3. Will quota, production targets, or incentives help to improve results?
4. Was it fair to punish or reward any worker? Why?
5. If you were the manager, what would you do to improve productivity and reduce costs?

When the syndicate groups respond to the first question, often there is no broad agreement as to whether or not the data are stable. This is a significant improvement from the mid-1980s when usually there was general agreement that the data were not stable. In order to create more uniform understanding, the audience then constructs a frequency distribution using data from two experiments, their own data and data from an earlier experiment. Data from two or more exercises are needed to

TABLE 5.2
Red Beads Experiment — Second Phase

Worker	Day 1 am	Day 1 pm	Day 2 am	Day 2 pm	Day 3 am	Day 3 pm	Day 4 am	Day 4 pm
James	8	6	4	12	7	10	5	2
Shane	4	7	5	8	7	3	6	1
Annabelle	3	2	6	8	8	5	10	5

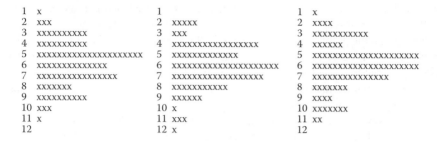

1	x	1		1	x
2	xxx	2	xxxxx	2	xxxx
3	xxxxxxxxxx	3	xxx	3	xxxxxxxxxx
4	xxxxxxxxxx	4	xxxxxxxxxxxxxxxx	4	xxxxxx
5	xxxxxxxxxxxxxxxxxxxxx	5	xxxxxxxxxxxxx	5	xxxxxxxxxxxxxxxxxxxx
6	xxxxxxxxxxxxxx	6	xxxxxxxxxxxxxxxxxxxxx	6	xxxxxxxxxxxxxxxxxxxx
7	xxxxxxxxxxxxxxxx	7	xxxxxxxxxxxxxxxxxx	7	xxxxxxxxxxxxxxx
8	xxxxxxx	8	xxxxxxxxxxx	8	xxxxxxx
9	xxxxxxxxxx	9	xxxxxx	9	xxxx
10	xxx	10	x	10	xxxxxxx
11	x	11	xxx	11	xx
12		12	x	12	

FIGURE 5.1 Distributions from red beads experiment.

obtain a distribution that is comprehensive enough for laypeople to understand. Then another one or two distributions are constructed, again using data from earlier experiments. Three distributions from one seminar are shown in Figure 5.1.

The audience is then asked to reconsider their assessment of the data. Almost without exception, they will conclude that the data are stable. They conclude that the variation exhibited is random but controlled, and that the results will fall between 0 and 12 with an average of about 5.7. They further conclude that until the beads, the method, or some other critical variable is changed, distributions will continue to look similar to those in Figure 5.1. Note that there are slight differences in the range, average, and shape of the distributions. This is natural and unavoidable because some element of sampling error is always present.

The next step is to plot the results of the first phase of the experiment as a run chart or time series plot. Such a chart is shown in Figure 5.2. If the data are stable, and they are, then Figure 5.2 shows what stable data looks like.

For many people, this is counterintuitive; but then again, new paradigms nearly always are. For many, their intuitive assessment of the data is that it behaves erratically. What the experiment does is help people to see order where previously they saw only chaos. The debrief includes the salient points that follow.

Nothing is changing. If the data are stable, they are trying to tell us that no significant changes are taking place.

Random variation. What confuses many is the statistical concept of predictability. Because all processes display some random variation due to common causes, it is true to say: "In a stable process, no one can predict what the value of the next datum will be, but anyone can predict the range, location, and shape of the next

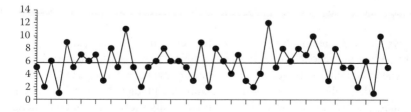

FIGURE 5.2 Plot of red beads data.

distribution." For the Red Beads Experiment, it can be predicted that the next result will fall somewhere between 0 and 12, and that is all that can be predicted. One can calculate that the probability of drawing a 5 is higher than the probability of drawing a 10, but it can never be known in advance when the next 5 or 10 is coming. It is never possible to predict what the next number will be in a stable system. The only way out is to reduce the range of the process outcomes; that is, to reduce the amount of random variation.

Same system, varying results. Because the data are stable, the same system that produced the high numbers produced the low numbers as well. The difference between them is random. Nothing changed.

Superstitious knowledge. In business, government, and education, the under-standing that two consecutive results can be quite different despite there being no change in the process is not common. In most cases, a cause is sought and acted upon. Imagine, however, that during the Red Beads Experiment, the manager phys-ically struck any worker who produced 9 or more red beads. Imagine also that the experiment was conducted not for 4 days, but for 400. What would such a manager learn? He would learn that in nearly every case, the next result produced by that worker would be an improvement, and that violence worked. However, he could have kissed the worker and proved that kissing improved results. Probability theory shows that the probability that two consecutive results will be 9 or higher is about 3%. Our lives are riddled with similar examples of superstitious learning. All of us have, in a like manner, developed erroneous cause-and-effect relationships. When these erroneous understandings are pointed out, it is common for us to react emotionally.

How do you know? It is common during management meetings to compare the current months' sales, safety, quality, and cost results to the previous month and to the budget. It is equally common for someone to then state that sales are up from last month, or that the accident rate fell this month. A relevant response is, "How do you know? Show me your plot." Any data must show some degree of variation, although such variation can nearly always be reduced. About half the data will be below average, regardless of the desirability or otherwise of this phenomenon.

HOW DO YOU KNOW?

A further experiment with the red beads data is to take the plot in Figure 5.2 and to project the center line into the future as a prediction of what will happen if nothing changes. It can be predicted that if nothing changes, additional points plotted will fall at random about this projected or predicted center line.[3] At seminars, the audience projects the center line for phase one of the exercise, and then plots the results achieved in phase two by the three remaining workers. Of course, the results have always fallen at random about the predicted center line because the process remains stable. What the second phase data is saying is: "Nothing has changed."

The center line is projected again, and the manager takes samples calling the results to the audience who plot them against the predicted center line. The job of the audience is to tell the manager when the results show a nonrandom change. At some point, usually after about 5 to 10 samples have been plotted, the manager subtracts two or three from each actual red bead count as he calls them out. The audience soon senses that something has changed. The evidence of this is that seven or eight consecutive results will fall below the projected center line. Something did change. This illustrates how one should answer the question, "How do you know whether the process has changed?"

IMPROVING THE ANALYSIS

In order to better understand and interpret a plot of points, control limits can be added. A couple of warnings are appropriate here:

1. It is a mistake to use the standard deviation function in your calculator or computer to calculate the standard deviation of the data, and then to measure two or three standard deviations either side of the center line to develop control limits. Although this method may work well when the data are stable, it fails utterly when they are not, and such limits can make even the most chaotic data look stable. Shewhart developed methods that largely overcome this problem, but it is noteworthy that many pharmaceutical companies have yet to include Shewhart's methods in their approaches to data analysis.

2. One of the most important things anyone does when analyzing a run or control chart is to establish the center line. Many computer packages will put a single center line through two or more systems of data. Most people are probably better off with a clear plastic ruler and a pencil when drawing in the center lines than they are if they let the computer do it for them. Figure 5.3 illustrates a poor analysis and a superior one.

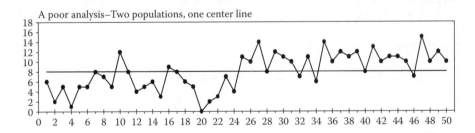

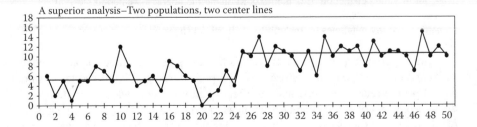

FIGURE 5.3 Two analyses of data.

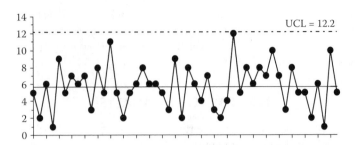

FIGURE 5.4 Control chart for red beads.

In Figure 5.4, the red beads data shown earlier have been used again. Here, the upper control limit (UCL) has been calculated and drawn on the chart. (In this case, the lower control limit [LCL] is zero.)

A control chart is the process talking to the process owner. In outline, this is what it says:

> Providing you do not change me (the process) or the inputs to me (raw material, training, methods of inspection and test, etc.), and providing only common causes of variation exist, I will continue to produce outcomes that fall at random between the LCL and UCL.

Moreover:

> Because the LCL and UCL are limits of controlled variation only, points falling at random beyond the control limits are a consequence of assignable (or special) causes. They have nothing to do with me (the process).

Therefore:

> If all points fall at random about the center line and within the control limits, I am stable.

Finally:

> If the degree of variation I display (when stable) and/or the location of my center line is not to your satisfaction, it is pointless to blame the operators. Operators can affect special causes only. Once I am free of assignable causes, only you (the process manager) can change my outcomes by making fundamental changes to me (the process).

One important aspect to bear in mind is that the process includes people. Better skills and knowledge, more uniform working procedures, and improved leadership are all people-related matters, and all affect quality in some way. Any attempt to distill people out of a process makes about as much sense as trying to distill a gallon of speed out of a sports car.[1] In many cases, human error can be traced to systematic causes.

Because the control limits are statistically calculated as limits of controlled variation, they must never be confused with targets or specifications. Control charts are based on an entirely different view of variation than are charts that show specifications. The basic difference between the traditional view of variation and Shewhart's view is:

The traditional view: Separate OK outcomes from failures by means of specifications or standards.

The Shewhart view: Separate processes that are stable from those that are not. Separate random from nonrandom variation.

These two views are very different and must never be mixed on one chart. A Shewhart control chart is used to determine stability. A chart with specifications is used to determine capability, or the ability of the process to produce OK outcomes. It is quite possible for a process to be producing stable garbage. The important understanding here is that if a process is stable, but producing outcomes that are not OK, the responsibility for improvement rests with management and technical people. The operators have nothing more to offer except improvement ideas.[1]

There is nothing new in these concepts. Shewhart's original work was done in the 1920s and 1930s. In 1953, during an address at the New York N.C.R. auditorium, Juran noted that when a worker achieves a state of statistical control she is contributing all she can. It was Juran who established the notion that approximately 85% of all variation will be built into the process or due to common causes. Only approximately 15% of all variation will be assignable or of a type upon which operators can sometimes successfully act.

DETECTING INSTABILITY USING CONTROL CHARTS

If a point falls at random beyond the control limits, as noted at Figure 5.5, a special cause has affected the process. This point is not random. Something has occurred to upset the process. The event that caused this point is not a function of the process itself; rather, it is special or assignable and should be investigated so corrective action can be taken and a recurrence prevented.

Figure 5.5 illustrates yet another type of statistical signal. Toward the end of the chart, eight consecutive points fall below the center line. This is statistically significant

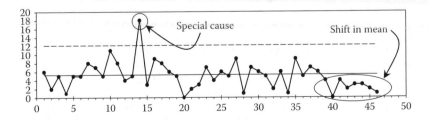

FIGURE 5.5 Detecting instability using control charts.

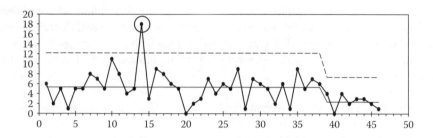

FIGURE 5.6 Redrawing the control charts for easy interpretation.

evidence that the process mean has changed. There are many other tests for stability, some of which will be covered in later chapters.[3]

Figure 5.6 shows this example recalculated and redrawn to better illustrate the shift in process mean. This makes it simple for even a layperson to correctly identify what the data are trying to say, making interpretation simple.

CHEMICAL EXAMPLE FROM THE PHARMACEUTICAL INDUSTRY

The chart in Figure 5.7 comes from a chemical process. (The scale has been omitted to protect the confidentiality of the data. Readers will note that the omission of scales is common throughout this book for the same reason.) The chart shows the concentration of the active pharmaceutical ingredient (API). This chart is stable, predictable. Only random variation within control limits exists. In turn, this means that no point is statistically high or low. The variation noted is random and constant over time. Given the number of variables that could destabilize a process, Shewhart was astounded at just how often complex manufacturing systems could be made to behave in as stable a manner as the Red Beads Experiment. This most desirable state was a consequence of several months of determined effort by the manager of this process, along with her technical people and operators.

Initially, the process was unstable. The manager formed a team consisting of herself, two chemists, an engineer, a laboratory analyst, and two operators.

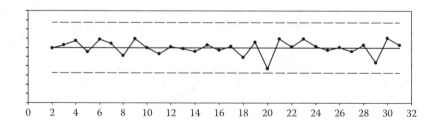

FIGURE 5.7 Concentration of the API in a chemical process.

They conducted retraining of operators and chemists to eliminate unnecessary adjustments, or over-control, which was evident from the initial charts. They improved the uniformity of sampling procedures, which were not always being followed scrupulously. Then they set about ensuring that the batch set-up was uniform from shift to shift, and that every batch was treated in the same repeatable manner regardless of which shift was operating. The process came into a reasonable state of control quite quickly, but was still too variable. This completed phase one of the project, to bring the process into a stable state.

In phase two, the improvement phase, the team investigated and found that at a distillation step the operator's instructions read, "Distill until complete." They discovered that operators were interpreting "complete" differently. An operational definition was created and all operators were trained in the definition and its application. Uniformity improved to the level noted in Figure 5.7.

BIOLOGICAL EXAMPLE FROM THE PHARMACEUTICAL INDUSTRY

The two charts in Figure 5.8 come from a chemical cleave step in a biologic process. The top chart shows the data with limits set at ±3 standard deviations, as is normal practice throughout the pharmaceutical industry. (Sometimes the limits will be set at ±2 standard deviations, but if this had been done in this example, the lessons would be similar). No points are observed beyond these three-sigma limits.

The lower chart shows control limits calculated and placed according to Shewhart's methods, and much more information is available. Those already familiar

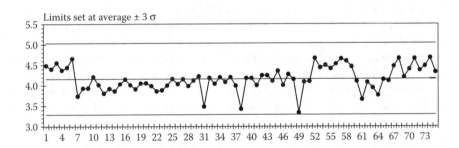

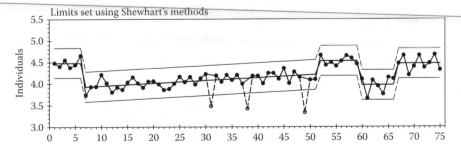

FIGURE 5.8 Two charts for a chemical cleave event.

with Shewhart's control charts will note that the moving range chart has been omitted for clarity, but it will be covered in detail in later chapters. The broken lines indicate that the point has been removed from the calculations.

The process in Figure 5.8 is unstable. It is not behaving in a predictable manner. The chart created using Shewhart's methods exposes special causes and several changes to the process mean. Later chapters will explain how such a chart may be created. For now, it is sufficient if readers observe that the Shewhart chart offers managers and technical people far more information about the process. It tells them where something changed the process average, and where upsets introduced special cause signals. The more traditional top chart found no signals in the data. The Shewhart control chart found seven.

COMPLIANCE EXAMPLE FROM THE PHARMACEUTICAL INDUSTRY

The chart in Figure 5.9 comes from a chromatography step in a plant producing a biologic product. What is noted is that there were two fundamental changes to the process mean and one special cause. The process is wandering up and down, but no one at the plant realized the process was behaving in such a manner because most outcomes were meeting specifications. This go, no-go attitude toward data analysis can be a major obstacle. The problem here is that batches 14 and 28 tested low enough to be noncompliant.

The chart in Figure 5.9 demonstrates that investigating batch 14 for reasons for noncompliance was the correct course of action. The point is special or "different" and warrants investigation. However, the investigation into the reasons for noncompliance at batch 28 was unsuccessful initially because only that batch was investigated. Batch 28 is not special; it is a random point on the control chart. The process drifted down at about batch 22, and this is the root cause of the failure of batch 28. After the process drifted down at batch 22, a noncompliant result was almost certain. Eventually a randomly low point would be low enough to be noncompliant.

If the managers and technical people at this plant had been using a Shewhart control chart, they would have noticed the change to the process before it became

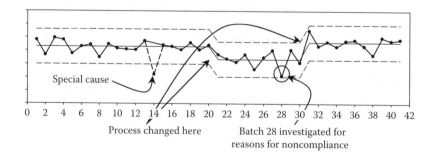

FIGURE 5.9 Control chart for a chromatography process.

noncompliant, and would have had the opportunity to take corrective action before a deviation occurred. Being proactive rather than reactive is an oft-quoted fundamental of sound management. Shewhart's methods can be of significant assistance to those managers and technical people wishing to control the work and be more proactive. The fact that so many plants in the pharmaceutical industry are plagued with deviations is evidence that much room for improvement exists.

THE ATTRIBUTES OF A BINARY MINDSET

In highly regulated industries, regulatory conformance is a critical business issue. The breaching of specifications, standards, or good manufacturing practice (GMP) brings grief to everyone. This has led to such a strong focus on specifications that tunnel vision has become common. Many people can think of nothing else but specifications and standards when they are asked to think about variation. Either the specifications were met or they were not. GMP was followed or it was not. The standards were achieved or they were not. These are important considerations, but they are outcomes only. Often, attempts to convince people to take action on a statistical signal are met with a shrug and a comment to the effect that specifications and standards have not been breached, so no action is required. Yet another Semmelweis Reflex occurs. As far as understanding a process and its variability are concerned, this type of binary thinking is the short road to hell.

ONE POINT LEARNING

1. Stabilize first *is* the first principle.
2. All managers and technical people should learn Shewhart's methods, so they can pursue stability.
3. Six Sigma focuses on creating predictable processes so managers can be proactive rather than reactive.
4. Causal relationships are easier to determine once the data are stable.
5. Meeting standards or specifications is *not* enough.
6. Shewhart's methods make investigating compliance issues easier.

REFERENCES

1. W.E. Deming, *Out of the Crisis*, MIT Press, Cambridge, MA, 1988.
2. W. Scherkenbach, *Deming's Road to Continual Improvement*, SPC Press, Knoxville, TN, 1991.
3. E.L. Grant and R.S. Leavenworth, *Statistical Quality Control*, McGraw-Hill, New York, 1980.

6 Action and Reaction

The Nelson Funnel Experiment

So what? Is all this stuff about variation really so important? What will happen to managers who do not adopt and use Shewhart's approach? Is there a financial penalty for ignorance?

The Nelson funnel experiment was devised to illustrate what will happen if people react to individual results in a stable process.[1,2] Unfortunately, the equipment it requires, a funnel, stand, and marble, makes it slow and cumbersome to execute in a classroom. What follows is an alternate method to teach the same lessons, devised during a control chart theory and practice course by a foreman at the Boyne Island Aluminium Smelter in Queensland. It is quick and effective and has the advantage of involving everyone in attendance at a course or seminar. To conduct the experiment, each person or syndicate group will need four sheets of paper and a felt tip pen. A small whiteboard marker works well.

THE NELSON FUNNEL (OR PEN DROPPING) EXPERIMENT

Nelson established that, in essence, there are only four ways to adjust or not adjust a process in order to compensate for error. Innumerable combinations and permeations exist, but there are only four basic underlying methods. Nelson called these the rules for adjustment of a process. The exercise is best conducted by commencing with Rule 4 and working backward to Rule 1. In outline, these rules can be summarized as follows:

Rule 4: Aim at the most recent result.
Rule 3: Adjust the aiming point by a vector equal to and opposite of the last error. Adjustment is measured from the target. (Error measured from target-adjustment measured from target.)
Rule 2: Adjust the aiming point by a vector equal to and opposite of the last error. Adjustment is measured from the last aiming point. (Error measured from target-adjustment measured from last aiming point.)
Rule 1: Aim every drop at the target.

RULE 4

Take the first sheet of paper and clearly mark a target in the middle of the page. Stand with the felt tip pen in your hand. Aim the pen carefully at the target holding

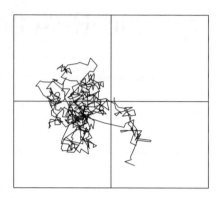

FIGURE 6.1 Results of Rule 4.

the pen from the very end so it hangs vertically. When the pen is aimed at the target, drop the pen. It will leave a distinct mark on the paper where it hits. Drop the pen again, this time aiming at the ink mark left by the pen's first impact. Continue to drop the pen, always aiming at the most recent pen mark or impact point. Draw lines that connect consecutive points. Drop the pen 50 times, unless the points are falling off the sheet of paper in fewer drops. A typical result using Rule 4 appears in Figure 6.1.

Rule 4 is used whenever the most recent result is used as the start point for the next activity and the activity is repeated. A common example occurs when someone makes a copy of a cartoon found in a magazine and gives it to a friend. This person makes another photocopy, using her copy as the master, and gives this new copy to an associate. Yet another copy of a copy is made and so on. After a while, the copies are unreadable.

When we were children, we played a game where a rumor was created and passed to a friend. The rumor circulated through the school, eventually finding its way back to its creator, usually changed to such an extent that it was hardly recognizable. The game had a variety of odd names such as "telephone." In business it is called "management information systems." Almost every senior manager has at some time complained about the difficulty experienced in obtaining accurate reports and good data from the cutting edge of the business. If the information must pass through many people before it can be acted upon, Rule 4 can play havoc with the original thrust and intent of the message.

Probably the most devastating example of Rule 4 is worker training worker, in succession. Suppose a business hires employee A and trains this person. Employee B is then hired, and is trained by A. As the business expands, employees C, D, and E are hired and trained by either A or B. Now A leaves to take a job in a new company and B takes maternity leave. New employees are trained by C or D, and these new employees are soon training others. In the long run, the level of on-the-job skills will decline. This phenomenon happens everywhere, but is easiest to see in new or high-growth businesses. It can be devastating.

A Pharmaceutical Example of Rule 4

During a plant tour, an operator was observed taking a sample from a large vat of broth containing cellular material in suspension. Such a broth is inherently difficult to sample without incurring significant sampling error. The procedure adopted was to take a triplicate sample. The broth was agitated for several minutes and a sample was taken with a ladle and transferred to a sample bottle. After a few more minutes of agitation, another sample was taken, and then after more agitation, a third sample was drawn. These three samples were tested and an average of the three was used as the test result for the vat. However, the observed operator actually took a sample with a ladle, which she then used to fill all three sample bottles. The entire purpose of the triplicate sample was destroyed. It was discovered that this operator was new to the job, and had been trained to do this by another operator, who had been trained by yet another operator. Rule 4 was being used to train the people taking the sample. There is a better way.

Quarterly reviews often use Rule 4, or a variant of it. If first quarter results are used as the plan for the next quarter and so on, Rule 4 is being applied. Where operational definitions and clear written and verbal instructions are absent, Rule 4 will step in. If sales figures from last month are used as the production plan for this month, Rule 4 may be being used, or perhaps some combination of Rule 3 and Rule 4.

As absurd as Rule 4 seems, it is the most commonly used approach. Most people tend to start any action, plan, or policy with the current reality. This encourages the use of Rule 4. Perhaps a better plan is to start every action, plan, or policy with a vision; an expression of which might be, "Where do we want to be?"

If Rule 4 is a random walk leading to somewhere, somehow, it does not sound like a good approach for business, government, and education.

RULE 3

With a fresh sheet of paper, set a new target. Drop the pen once. Draw a freehand line from the impact point back to the target, then continue that line an equal distance past the target and set a new aiming point at the end of this line. So, if the first impact point was 14 mm high, the new aiming point would be 14 mm below the original target. Drop the pen 50 times, setting the aiming point equal to and opposite of the target on each occasion. An illustration of Rule 3 is shown in Figure 6.2.

As the experiment is conducted, participants note that both the aiming and impact points eventually start to swing, pendulum style, across the original target. Rule 3 occurs whenever the process is adjusted by an amount that is equal to and opposite of the most recent error.

A Pharmaceutical Example of Rule 3

In Figure 6.3 is a chart that shows a period of over-control by an automatic process controller. The controller is attempting to maintain a certain level of concentration of a protein in a solution. Points that are either high or low enough to engage the

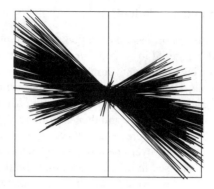

FIGURE 6.2 Results of Rule 3.

controller can result in over-control, which often is characterized by a saw-tooth pattern. This pattern is circled in Figure 6.3. Control limits are not necessary here. The pattern and shape of the data tell us all that is necessary to know.

In this case, the controller was switched off and the over-control ceased immediately. Reprogramming of the controller was necessary so it could more adequately discriminate between random and non-random variation. This all but eliminated any further over-control.

Control instruments are common in the pharmaceutical industry. It is the responsibility of managers and technical people to ensure that they are programmed in such a way that they properly do the job for which they are designed. Automatic controllers can both over- and under-control, but the former is much more common. Before control engineers can adequately test the programs for these controllers, they will first need an understanding of Shewhart's statistical methods.

Financial control and inventory control are two areas where Rule 3 is often found. Whenever there is "feast followed by famine," suspect the presence of Rule 3.

Possibly the worst example of Rule 3 can be found in production planning and scheduling. If the sales figures are reasonably stable, the worst thing managers can do is react to the individual data. Factory managers often report that it is impossible to run an efficient operation when the production schedule is jerked up and down.

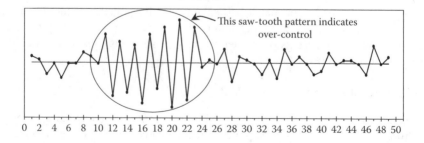

FIGURE 6.3 Over-control of the concentration of a protein.

Continuous and semi-continuous operations suffer the most when Rule 3 is applied to production scheduling. Variation is not free. It is expensive. A good place to start is production planning and scheduling. If sales are stable, aim to produce to the center line every production cycle. There is a big difference between being market led and chasing random variation.

In a similar vein, it can be a mistake to react to the most recent sales figures, especially if sales are reasonably stable. One common problem occurs when a randomly low point or two produces the reaction of conducting a sales promotion or dropping prices to move product. If the sales result is random, then the chances are that the next result will be higher even if no action is taken.

Some companies are learning that their monthly reaction to sales, volume, and profitability is destructive, and they are finding better and safer ways to react. If the newspaper reports are correct, it appears that Proctor and Gamble is one such company.

DISCOUNTS DISCOUNTED

Last year, when the Proctor & Gamble Co. introduced its new pricing strategy of eliminating discounts on its consumer products in favour of consistently low prices, analysts were dubious. But last week Edwin Artzt, the company's chairman, told shareholders the strategy was paying off in sharply reduced costs. Proctor & Gamble expects savings this year of $US174 million from eliminating incentives. 'Without discounts, demand for products is more constant so the company's factories can operate more efficiently,' Mr Artzt said.[3]

It appears that Proctor & Gamble discovered that certain marketing actions were in fact increasing variation in sales. Much of what is done in an attempt to "manage" variation in sales can actually make matters worse. Sales and marketing people have a new facet to their jobs: variation reduction. Many managers are so action-orientated that to ignore a low result and maintain focus on the existing strategy seems absurd; yet, this is a far better approach if the data are stable. The trick is to separate "signals" that indicate something is changing from the "noise" of random variation. Shewhart's control charts are the best method yet devised for this job. They apply as much to sales, marketing, and distribution as they do to production. This is easy to say, but it does not make the job of the person trying to convince the sales and marketing people to use statistical tools any easier, unless the business has made reducing variation a strategic imperative.

The other major problem with sales data is that if a randomly low point is followed by a higher one, people can convince themselves that the actions taken to increase sales actually worked. Maybe they did, and maybe they did not. The only way to be reasonably sure is to plot the sales data as a control chart. If the data exhibit stability, then these data are trying to demonstrate that nothing is changing. The sales team may have exerted great effort, but the key causes of sales volume remain unchanged.

RULE 2

Start with an aiming point and a new sheet of paper. Drop the pen once, aiming at the target. Because the aiming point and the target are coincident for the first

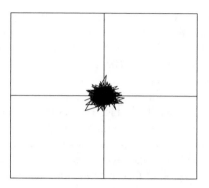

FIGURE 6.4 Results of Rule 2.

drop, the first adjustment is as per Rule 3. Drop the pen again, aiming at the new aiming point. Draw a freehand line from the impact point back to the target. Starting at the most recent aiming point, draw a second line that is equal and parallel to the first line. If the second drop were 20 mm left of the target, the next aiming point would be 20 mm to the right of the last aiming point. Drop the pen 50 times, again connecting consecutive points. An illustration of Rule 2 is shown in Figure 6.4.

Rule 2 is stable, but it shows twice the variance (σ^2) of Rule 1. For discussion purposes, Rule 2 is similar to Rule 3, so no great detail will be entered into here. However, there are two important points worth noting:

1. If Rule 2 is being used, one of the few ways to expose it is to conduct a trial and allow the process to run with no manual or automatic adjustments for a while, carefully monitoring the degree of variation. If the variation decreases, unnecessary reaction is built into the controller's logic.
2. Many automatic process controllers use a variant of Rule 2. The authors have yet to find a plant where at least one of the automatic controllers was not increasing variation. Finding these controllers and either replacing or reprogramming them is a quick way to improve quality, productivity, and profits. Questions that should always be asked of production and control engineers are: "How do you know the controllers are not over-controlling? By what method was your conclusion drawn? Can you show the trial data?"

A Pharmaceutical Example of Rule 2

The results of a trial to test the automatic process controllers in the fermentation event of a pharmaceutical plant are shown in Figure 6.5. The variable plotted is potency.

To conduct this trial, the automatic process controllers had a significant dead band created so the controller would only react in the event of a large change in the

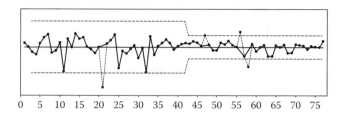

FIGURE 6.5 Rule 2 in a pharmaceutical plant.

process. Essentially, the controller was not used during the trial. Variation dropped immediately.

When Rule 3 is being used, usually the pattern of the data will reveal the problem. However, Rule 2 produces no such oscillating pattern and it will be necessary to test the controllers with a trial. This is not a difficult exercise, and it can produce some spectacular results.

Another common example of Rule 2 is the use of drug substance potency stability data to determine the amount of material needed for formulation. When the drug substance potency data is stable (in a state of statistical control and not increasing/ decreasing as a function of time), no adjustment from the last target is needed. Under these circumstances, adjustment leads to increased variability in the formulation process. Each formulation should be performed using the same original value, provided the data are stable. However, it is common to adjust the formulation parameters based on the latest value. In this way, the analytical variability is directly increasing the variability in formulation.

In a like manner, some continuous processes have their variation increased by unnecessary calibration of automatic controllers. In one example, the controller was doing an excellent job of holding the chemical composition of a blend of certain materials to its set point. The problem was the calibration process. Once per shift, a physical sample was taken. The analytical result was keyed into the computer, which calibrated the controller as if the laboratory result were perfect, free from sampling and analytical error. This procedure increased the variation by approximately 40%. A better way is to plot the laboratory results. If they remain stable about the target value, no calibration is necessary. If a statistically significant signal is recorded, a recalibration process should be enacted.

Finally, having concluded that Rules 2 to 4, or any combinations of them, inevitably increase variation, attention is drawn to Rule 1.

RULE 1

Prepare a new sheet with its target. Aim every drop at the target. Regardless of the results, do not adjust the way you hold, aim, or drop the pen. Conduct 50 drops. Figure 6.6 illustrates Rule 1.

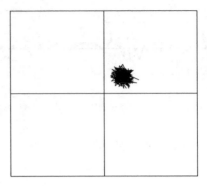

FIGURE 6.6 Results of Rule 1.

RESULTS OF THE EXERCISE

In Figure 6.6, an example of Rule 1 was chosen where the stable result is falling away from the target. In this case, an adjustment is necessary to bring the pattern of results onto target. If the reader understands the difference between reacting to individual points and deliberately adjusting a stable system of points, the major lesson of this experiment is understood, and may be summarized as follows:

> It is not possible to reduce variation in a stable system by adjustments based on the most recent results

Or:

> If your routine is to take action based on the most recent results, variation can only increase.

There is nothing new in any of this. This understanding of variation is taught to every soldier as the theory of the group. In order to improve any process, a simple procedure that works is the following:

1. First, study the data for stability. If the process is not stable, find and eliminate the assignable causes. As far as possible, eliminate the variation between operators and between shifts. Achieve repeatability.
2. Once the data exhibit stability, start at the upstream end of the process and look for ways to reduce variation. A process-based team is best for such a study.
3. When the variation has been reduced, adjust or shift the process onto the desired value.

The approach is a far cry from reacting to individual results and it works. However, what kind of reaction would be expected from most operators if they were told that it could be a fundamental error to adjust a process because the latest outcome was out of specification, that stable garbage was far better than chaos? If the finance director is told that the 14% drop in profit last month was random and that she need not be concerned about it as yet, what kind of reaction would be anticipated? What would happen if the sales manager was informed that the 8% rise in sales last month had less to do with the latest promotional activities than it did with random variation?

Our mental models or paradigms about variation are, in most cases, very different from the concepts developed by Shewhart. There is neither right nor wrong in our world, not until the subject at hand is couched in terms of our knowledge. Those who choose to use the approaches pioneered by Shewhart can expect significant progress. Those who choose to ignore Shewhart's approach, or remain unaware of it, will find improvement problematic.

One area of concern is the practice of adjusting a process based on specification limits. In processes that are significantly more variable than Six Sigma levels of performance, adjustments based on specifications will often lead to many unnecessary adjustments. A better option is to use a control chart with projected limits for process control, as discussed in Chapter 16.

SERVICE ELEMENTS OF THE PHARMACEUTICAL INDUSTRY

A common feature of service elements, and in particular those in mature markets, is that processes are mostly stable and robust. Usually, sales data for mature products are good examples of this inherit robustness, which is both good and bad. It is good inasmuch as when managers first run control charts with their data, the analysis is simple because service elements tend toward stability. The process is easy to "see." However, the downside is that quite often the inherent robustness means that significant or profound change is necessary to bring about any worthwhile improvement. Tinkering or fine-tuning seldom has much impact.

The evidence of this is to plot out the historical data and examine it. Isolate several key processes that look reasonably stable. Some examples are:

1. sales by product line
2. customer service levels
3. response time
4. costs of any kind

Then, list the "corrective action" taken at various times when results have been below average. Naturally enough, approximately half the data always falls below average. If the data are stable, this indicates that whatever action was taken, it had little or no noticeable effect on results. In these cases, it may be that the effect of reacting to individual data is not so much to increase variation, as it is to prevent improvement. This happens in several ways, two of which are:

1. **Superstitious knowledge**. In a stable system, a quite low result will nearly always be followed by a higher point, regardless of either the presence

or absence of a knee-jerk reaction. Often, this leads to the development of cause-and-effect relationships that are erroneous but believed to be true.

2. **De-focusing managers**. Imagine that a business has 10 key performance indicators and that all are reasonably stable. Any given monthly report will show approximately half above average and approximately half below average. Generally, one or two will be quite high, and one or two will be low enough to cause concern. This leads to a focus on seemingly poor results, even when they are random. Next month's report shows these results to have improved, but another one or two key indicators have taken their place as areas for concern and action. The obvious resultant is that every monthly report brings with it a new set of priorities. This fire fighting prevents staff from focusing on a key problem long enough to create a breakthrough. Some problems take more than a month to overcome.

These problems can be found in any segment of the pharmaceutical industry, but are most obvious where the process is robust and mature. A robust process usually takes more time and effort to improve, yet the practice of chasing monthly variances prevents managers from taking the time they so often need to work on one or two key issues to achieve a breakthrough.

ONE POINT LEARNING

1. Stabilize first *is* the first principle. Sometimes this means we must learn to stop injecting unnecessary variation.
2. All managers and technical people must learn Shewhart's methods, so they can pursue stability and eliminate over-control. Stopping and then preventing over-control is an early imperative in attempts to reduce variation to Six Sigma levels.
3. Over-control exists in sales, scheduling, and all other areas. It is a mistake to focus on manufacturing only.
4. A process should *never* be adjusted based on specifications. Control charts should be used for this purpose.

REFERENCES

1. W.E. Deming, *Out of the Crisis*, MIT Press, Cambridge, MA, 1988.
2. W. Scherkenbach, *Deming's Road to Continual Improvement*, SPC Press, Knoxville, TN, 1991.
3. *Australian Financial Review*, October 19, 1992, p. 10.

7 Close Enough; …Or On Target?

The Taguchi Loss Function

Dr. Genichi Taguchi provided a new mathematical framework to demonstrate that reducing variation around an optimum could be the correct economic choice, even when the product routinely met specifications. In 1960, Dr Taguchi won the Deming medal for his breakthrough.

Dr. Deming was often quoted as saying that it was good management to reduce the variation in any quality characteristic, whether this characteristic was in a state of control or not, and even if few or no defectives are being produced.[1] The Six Sigma concept introduced in Chapter 2 demands that the variability of any event in the process ought to be no more than half that allowed by the specifications. Many people found a requirement for such demanding precision difficult to understand, let alone believe. The mindset in most pharmaceutical companies is that if the characteristic meets specification or standard, no further improvement is necessary or warranted. Any occurrence outside of a specification or standard is treated as a deviation.

From Chapter 4 the second definition of quality was:

All characteristics must remain stable and within the go, no-go specifications.

Both Deming's approach and the Six Sigma methodology are at odds with this definition. Both are much more demanding. Taguchi was able to demonstrate that reducing variation far below specifications was the correct course of action. So, too, was the founder of Six Sigma, Bill Smith.

In the 1920s, Shewhart noted that quality was essentially an economic problem. Taguchi followed this line of thinking. He defined the cost of poor quality as the total loss incurred by society due to variation and poor quality. Taguchi was passionate about quality, to the point where he claimed that the manufacturer of poor quality was worse than a thief. If a thief steals $100 from a neighbor, he has gained and the neighbor has lost, but the net economic impact on society is nil. That $100 will still be invested or spent on goods and services. However, if a manufacturer throws away $100 in rejects and rework, the cost of wasted resources and to raise and clear unnecessary deviations can never be recovered by either the company or society.

One way to introduce the Taguchi Loss Function is to start with the notion that design engineers, chemists, and biologists do not like variation. They prefer perfect

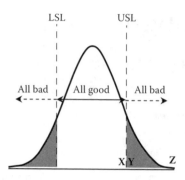

The old view states that **X** is a good result, and that **Y** and **Z** are equally bad.

However, in terms of their physical properties, X and Y may be nearly identical.

Is there a better way to describe their varying degrees of 'goodness' or 'badness'?

FIGURE 7.1 The old view of variation.

precision and have an optimum value in mind. However, they understand that perfection is impossible and so they create tolerances to delineate the boundaries of permissible variation. Nevertheless, there is always an optimum value (OV). For the sake of simplicity and clarity, those characteristics where bigger is always better (horsepower of a racing car) and smaller is always better (weight of a racing car) will be ignored.

Second, consider product X whose measured characteristic falls just barely inside the upper specification, as noted in Figure 7.1. Now imagine product Y, which is only just outside the upper specification. In terms of their physical properties and their performance in the customer's hands, these two products are likely to be, for all practical purposes, identical. In most cases, the sampling and analytical error involved will be greater than the observed difference between these two outcomes. Yet the methodology used in the pharmaceutical industry will always state that product Y is outside specifications and is therefore a deviation. Now let us add one more product to the chart, product Z, which is well outside the specification. The old "go, no-go" definition of quality states that all products outside specification are equally bad.

As noted in Figure 7.2, Taguchi envisioned a loss function curve, usually a parabola, which described the increase in loss on a continuum as outcomes moved away from the optimum value, rather than as a go, no-go situation. The vertical height from the baseline (no loss) to the curve described how the amount of loss increased as outcomes move further away from the optimum value, until complete loss occurs.[2]

Taguchi's approach made intuitive sense. Suppose one is using the go, no-go approach and the specifications for pH for a certain process parameter are 6.5 to 7.5. A reading of 7.45 will be accepted as in specification, but a reading of 7.55 will be out of specification, even though there is no discernable change in the product. The Taguchi approach sees these readings on a continuum. The target or optimum value is 7.0. Our loss increases continually the further one departs from this optimum. Absolute limits are still necessary for release and compliance reasons, but if loss is seen on a continuum, it opens new perspectives and opportunities to understand and improve the process and product hitherto unnoticed.

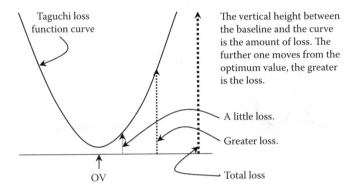

FIGURE 7.2 The Taguchi view of variation.

The Six Sigma approach demands abandonment of reliance on the go, no-go specifications. They are still necessary to set absolute limits, but should not be the sole basis for assessing the variation at any stage of the process. In addition, go, no-go specifications should never be used as the basis for process control, a concept that will be dealt with in detail in Chapter 16.

The Taguchi loss function takes several forms. One simplified form of calculation is shown in Equation 7.1 and Equation 7.2.

For any given product:

$$Loss_x = k(x - OV)^2 \tag{7.1}$$

where:
x = value of quality characteristic
OV = optimum value
k = a proportionality constant (a function of the failure cost structure)

For any given population of products:

$$Loss_x = k(\sigma_x^2 + (Ave_x - OV)^2) \tag{7.2}$$

When the Taguchi Loss Function Curve is overlaid on a distribution, the area of intersect represents the amount of loss, which can be calculated from Equation 7.2. Some loss function curves are low and flat, allowing considerable movement away from the optimum value before significant loss is incurred. Others are tall and steep, where even a small deviation from the optimum value can incur considerable loss, as noted in Figure 7.3. The steepness of this curve is driven by the size of the k factor.

In most cases, working on the earlier steps in the process, with particular regard to inputs and set-up of the process, provides significant leverage because the loss function curves are nearly always steeper there. The importance of working upstream in the process cannot be overstated. Too often, a plant's most talented people can

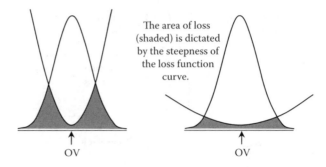

The area of loss (shaded) is dictated by the steepness of the loss function curve.

FIGURE 7.3 Finding the area of loss.

be found working in the middle or latter stages of the process, after variation in the earlier events has already destabilized the process and shrouded cause-and-effect relationships. Other statistical methods such as design of experiment (DOE) will help to isolate those variables that have the greatest impact on quality, deviations, and costs. A cornucopia of advanced statistical techniques exists, but it is beyond the scope of this book to cover them. For those seeking more detail on Taguchi's Loss Function and on industrial experiments, Reference 3 is a good starting point.

Taguchi's approach illustrates two important and interlinked points. First, it is in concert with Deming's statement that merely meeting specifications is an inadequate approach and that businesses should always be attempting to reduce variation. Second, it demonstrates that as variation is reduced, so too is the amount of loss incurred. It only remains to isolate those variables that provide greatest leverage. These concepts are illustrated in Figure 7.4.

Taguchi's concepts led to the development of the third definition of quality:

All characteristics should have minimum, stable variation around an optimum value.

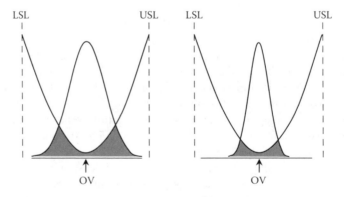

FIGURE 7.4 Reducing variation reduces the area of loss.

This is in concert with the Six Sigma concept outlined in Chapter 2. Six Sigma states that the nearest specification should be at least six sigma from the process mean. Taguchi's concept is another way to illustrate how variation affects processes and costs. What is clear is that several different approaches to the subject arrive at the same conclusions, as follows. First, variation costs money, even when everything appears to meet specifications. Second, reducing variation always reduces costs and deviations. Third, not all variables have an equal impact on costs and deviations. Some are very significant; others are not. In no small way, a critical part of the Six Sigma concept is to discover which variables harbor the greatest potential to reduce costs and deviations.

Chapter 8 introduces Little's Law, which explains how reduced variation in the flow of material through a process also increases throughput volume, especially when the earliest events in the process are addressed. Variability affects both quality and quantity.

Although the loss function curve can be calculated for any variable, it is not necessary to do this in order to grasp the core concepts and to make a start. If it is understood that the early steps in the process generally have steeper loss function curves for both quality and quantity, then work to reduce variation can commence there immediately. The flow-on benefits throughout the process can be dramatic.

These understandings of variation and the reduction of variation are critical in the pharmaceutical industry, where plants must raise and clear a deviation for every outcome that fails to meet specification or standard, or when the approved process had not been followed. This consumes resources, money, and technical people's time. In extreme cases, high variation can lead to a consent decree. Where a significant proportion of technical people's time is expended raising and clearing deviations, time is wasted that could have been spent improving the process and the product. Reducing variation by achieving a stable state and then minimizing variability is crucial in regulated industries. Close enough is not good enough for the pharmaceutical industry.

ONE POINT LEARNING

1. Meeting specifications is not good enough.
2. Loss is incurred even when all outcomes meet specifications.
3. Managers and technical people need to know which variables have the steepest loss function curves, and to start work there.
4. Minimum, on target, stable variation that places the specifications at least six sigma from the process mean is the standard.

REFERENCES

1. W.E. Deming, *Out of the Crisis*, MIT Press, Cambridge, MA, 1988.
2. W. Scherkenbach, *The Deming Route to Quality and Productivity*, CEEPress, Washington, D.C., 1988.
3. D. Wheeler, *Understanding Industrial Experimentation*, 2nd ed., SPC Press, Knoxville, TN, 1990.

8 Make More...Faster!

Little's Law and the Dice Experiment

Often one can find managers trying to strike a balance between quality and quantity. Such a balance does not exist if the definition of quality includes low variation. We can have both.

There has been a significant increase in quality and productivity improvement methodologies in the past 25 years. However, to this day there are those who call into question the wisdom of focusing on quality because of fears of lowered production rates. In service as well as in manufacturing, unit costs are important, and unit costs are heavily impacted by the volume of work done by any given mix of people and machines. In addition, cycle time and speedy reaction to the market are important considerations. This chapter addresses these issues and demonstrates the links between variation and both volume capacity and cycle time.[1]

THE DICE EXPERIMENT

A simple experiment with playing dice helps to illustrate the important links between quality, variation, output volume, and work in progress. The experiment is set up as follows:[2]

1. A supplier, seven department managers, and a customer are recruited from the audience.
2. The supplier, managers, and customer are formed into a process, where raw materials or information is fed into the process by the supplier, and where each manager passes the job to the next manager once his or her actions on the job are complete. Put another way, they form a series of dependent events, where each step is dependent upon the earlier steps.
3. The supplier has a supply of "widgits" (usually books or similar items) that represents incoming materials, goods, information, or paperwork.
4. The customer uses an overhead projector or whiteboard to record the volume of work that reaches him or her each day. The customer is interested only in volume. Quality and cost are not important in this exercise.

A scenario is set whereby customer research has determined that the customer's requirement is 35 widgits per 10 days, with an average of 3.5 widgits per day. A playing die is then produced. It is a stable system (although it does not produce a normal distribution) with an average of 3.5. This die will be used to simulate the

TABLE 8.1
Typical Results — The Dice Experiment

	Daily Volume Output to Customer		
Day	Die No. 1 (Avg = 3.5)	Die No. 2 (Avg = 4.5)	Die No. 3 (Avg = 3.5)
1	1	2	3
2	1	3	3
3	1	2	3
4	2	4	3
5	4	2	3
6	1	2	3
7	4	6	4
8	1	3	3
9	2	2	3
10	4	3	3
Total	21	29	31
Daily average	2.1	2.9	3.1

output for each step in the process. There will always be some degree of variation about the average, so on some days more than average volume will be expected and on other days lower than average volume would be anticipated. Because each step in the process uses the same die to simulate daily output, each step in the process will have the same degree of variation and the same long-term process average.

The exercise commences. The die is thrown for the first step, which is the supplier. Whatever number appears on the die is the number of widgits passed on to manager 1. The die is then thrown for manager 1, who passes on to manager 2 the number of widgits that corresponds to the number shown on the die, and so on to the customer who records the number of widgits that actually arrived on that day. Sometimes, a manager will throw a number that is lower than the number of widgits held. After passing the required number of widgits to the next manager, the widgits still held represent work in progress. When a manager throws a number greater than the number of widgits held by a given step in the process, the situation mimics those cases where more work could have been done, but where insufficient feed was available from previous steps.

At the end of the 10-day's work, the volume that reached the customer will be much less than 35 widgits. Often it is less than 20. In addition, work in progress or inventory is high, resulting in high carrying costs. A typical result, such as has been obtained after many exercises, is shown in Table 8.1, under Die No.1.

A common response from those pressured to increase production is to ask for more resources. This situation can also be illustrated during the experiment by using a die that has had the side that normally shows a 1 converted to a 7. The new die varies from 2 to 7 with an average of 4.5. Nearly 30% more resources have been added. Using the same supplier, managers, and customer, the exercise is repeated with the new die. A typical response is shown in Table 8.1 under Die No.2.

One would expect the volume of work reaching the customer to be greater when using the second die, and it is. However, bear in mind that significant additional resources and, therefore, costs were expended to increase volume to this new level. Work in progress remains at a high level, as do carrying costs.

At this stage of the exercise, yet another die is produced. The audience is told that this is a quality die, not a volume die. The audience is told this die, like the others, is a stable system. It has a process average of 3.5. The extra resources used for the second stage of the exercise have been removed. The audience is asked to watch the next stage of the experiment and determine what makes the latest die a quality die.

After only a few throws, the audience will have figured out that the quality die has been altered from its original state; its six faces show only two possible outcomes; three sides show a 3 and three sides show a 4. The exercise continues until 10 days of work have been done. The customer's record shows that the quality die produced the better result in all respects:

1. Volume is up.
2. Work in progress is down.
3. Cost per unit is down.
4. Time lapse between customer demand and customer satisfaction is shorter.
5. Volume output is more predictable.

An inescapable lesson from this exercise is that if the objective is to create greater volume output, then businesses would do well to learn about how variation affects not only quality, but also of quantity. Six Sigma levels of variability improve both quality and quantity. Little's Law explains the connection between variation and volume output. It demonstrates that in any series of dependent events, the volume output is dependent upon not only the average output of each event, but also its variation. Work processes are a series of dependent events. That is, a step in the process cannot be completed until that work has been passed on from an earlier step. In this way, any step in the process is dependent upon the earlier steps in terms of quantity as well as quality.

LITTLE'S LAW

The Dice Experiment is a practical demonstration of Little's Law, named after Dr. John Little[1] of MIT who was awarded the world's first Ph.D. in operations research. He provided the mathematical proof of this law. Little's Law is expressed as the formula:

$$Throughput\ Volume = \frac{Work\ In\ Progress}{Cycle\ Time}$$

There is a corollary to this law, which states that as variation in the flow of material or information decreases, so too does cycle time. Assume variation and cycle time are reduced. For the equation to remain balanced, work in progress must fall, throughput volume must increase, or some combination of the two. Therefore, reducing variation either lifts output or reduces work in progress, or both.

As one might expect, increasing variation has the opposite effect. In addition, operations research has demonstrated that reducing variation in the earlier steps of a process has a greater impact than does reducing variation in later steps.[1,2] This is as one would intuitively expect matters to unfold. Variation in the early steps affects all downstream events, whereas variation in a much later event can affect fewer events.

There are several ways of increasing volume through a process without large increases in inventory or resources. Two will be discussed here.

Reduce Variation. The type of variation referred to in Little's Law is variation in the flow of work. If the entire process can be made to operate with very low levels of variation in throughput at each step, significant benefits will follow. Reduced variation lowers cycle time. This then gives the process owners the option either to drive down work in progress or to lift throughput volume of the process. This is a straightforward business decision.

Reduce the Number of Steps. The principle of "Keep It Simple" is a popular part of our folklore. If the experiment with the dice were to be repeated with 50 steps, work in progress would soar and volume output would fall. The opposite affect would be noted if the number of steps was reduced. In this experiment, each manager represents one department in the process. In practice, it is the number of events in the process that is critical. An event is a part, raw material, or processing step; it is any opportunity for something to go wrong. Motorola calculated that there are 5000 events in the manufacture of a cell phone.

Most businesses can find examples of high variation choking volume and causing problems in the latter part of the process. One example is the surge in production or shipping that so often happens toward the end of a month or quarter. In some companies, a plot of stock movement or production shows a very high point at the end of each of these reporting periods. These bursts of activity only increase variation and either damage the downstream process or result in inefficient use of resources. Instead of running through the process at a steady rate, product is rushed through in lumps. Someone is paying for this practice.

This situation is not uncommon. In some businesses, it is routine. The job of managers in some companies is not to optimize the entire process. It is to meet monthly or quarterly functional targets, with little or no regard for the downstream steps in the process. In such companies some managers can make their department look good, but at the expense of others, and, more often than not, at the expense of the customer and company profits.

How does a company set about reducing variation and reducing the number of steps in the process? In many cases, two courses of action achieve both aims. First, work now and forever to eliminate yield loss, rework, down time (including human down time), complexities, unnecessary bureaucracy, and any form of waste. This is the great irony of the quality vs. quantity argument. In most processes, process disturbances, rework, and down time are chief offenders among causes of variation in volume. Second, work to reduce variation in inputs, whether those inputs are components, raw material, or information.

It is particularly important that production planning and scheduling be conducted in such a way that the impact of variation is minimized. Six Sigma demands the reduction of variation in every step performed throughout the process.

Readers may have noticed that this lack of variation is what makes the much celebrated kanban or just-in-time system possible. There is a lot of material in print about just-in-time that fails to recognize that good quality, or lack of variation, is a key foundation for such an approach. Surveys of businesses that use Manufacturing Resource Planning (MRP) find that some companies swear by it, while others swear at it. It is noteworthy that the businesses struggling with MRP are nearly always those with higher variation.

The dice experiment can be repeated with many variations on the same basic theme. One such variation is to start each step with some work in progress. If this is done, things look better for a while but before long, the inescapable effects of variation take their toll. As the die continues to roll, the long-term effect is a gradual rise in work in progress, and volume output slowly sinks toward the smallest number on the die, or toward a theoretical limit. When the work in progress builds to a high enough level in the last couple of steps, large volumes of work will exit the process at high cost, in terms of both dollars and speed of satisfaction of customer demand. Soon the effect of variation will lower output and raise work in progress in an erratic manner; and so the cycle continues. Feast is followed by famine.

Finally, the Dice Experiment illustrates how important it is to manage a process as an entity. It appears paradoxical that the large vertically integrated companies that have the greatest opportunity to manage the process as a whole have chosen to break work processes into vertical functional lumps for management.

Most managers will agree that cross-departmental cooperation is necessary to optimize any process. What many do not grasp is that goals and incentives based on a functional approach will cause some managers to focus on the boss and the reporting system rather than on their downstream customers. Where cooperation is present, sometimes it exists despite the management system rather than because of it.

On a larger scale, businesses are beginning to understand that their suppliers are part of the process. For a long time, companies were intent on reducing costs of purchase when buying, or maximizing the volume or price tag when selling. Few were interested in banding together to drive down the total cost and making the product and service of maximum usefulness to the customer and growing the market. Now supplier partnering is becoming more common.

An army of supplier auditors exists in the pharmaceutical industry, but as is the case with most of the industry, they are almost exclusively focused on compliance and the meeting of standards and specifications. If auditors were trained in Shewhart's methods and the Six Sigma approach, they could help suppliers drive down variation. In turn, this would provide lower variability in inputs to the pharmaceutical manufacturing processes. It is not suggested that this would replace their existing function, but is an important amplification with a potentially profound impact.

QUALITY CONTROL CONSIDERATIONS

Imagine two identical processes. One has low cycle times and work in progress. The other has high cycle times and work in progress. Imagine that in both cases final inspection finds a problem with the finished product. In the business with high work in progress, the amount of production that is potentially affected by the problem and which must be quarantined and reinspected will be much higher than in the process with low work in progress. In the pharmaceutical industry, this can be a vital consideration. Rapid cycle time and low work in progress can be an important part of quality control and assurance activities.

SIX SIGMA AND FIRST PASS YIELD

Another way to illustrate the impact of variation is with the model used to explain the Six Sigma concept within Motorola in the mid 1980s, presented in Chapter 2. Given that yield loss and rework are chief among the causes of variation, one would expect to find high volume throughput only where first pass yield is high and rework is low.

The figure that illustrated three and six sigma systems in Chapter 2 is reproduced in Figure 8.1.

Assume that the process involved has 200 events, or opportunities, for something to go wrong. For the sake of convenience, correctly centered normal distributions

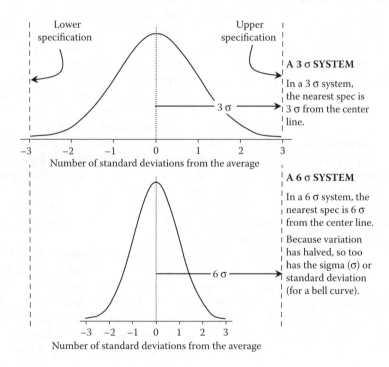

FIGURE 8.1. A three sigma system event and a six sigma system event.

are assumed. In the three sigma system every event will have 99.7% yield, in accordance with normal law. A perfectly stable three sigma system with 200 events will provide a first pass yield of only 58%, but in the Six Sigma system it will be 99.99%, as noted in Chapter 2. With low variation and the attendant high yields of a Six Sigma system, retests, deviations, and quarantined batches will be all but non-existent. In this situation, variation in volume or the flow of work will be equally low, and high throughput with low work in progress will follow.

In his early seminars, Harry emphasized that the Six Sigma approach was based on reducing variation. He described two approaches:[3]

1. **The shotgun approach.** This entails the creation of a culture where everyone in the business has the reduction of variation as part of his or her job. This is just the way things are done around here.
2. **The sniper rifle approach.** Formal projects isolate and improve those variables that give greatest leverage. The Dice Experiment assumes that every event has the same average capability and the same degree of stable variation. This is not an adequate representation of any plant. Some events have greater impact than others do, and statistical tools exist to help us isolate and improve these events.

There is much more to the Six Sigma approach than this simple concept, but reducing variation is at the very heart of the subject.

PHARMACEUTICAL CASE STUDY — INCREASING OUTPUT

A plant manager producing consumer healthcare products was struggling to meet demand. Quarantined batches, deviations, and downtime plagued the plant. Line by line, he set about improving maintenance, machine changeover and set-up, as well as compliance to standard operating procedures. Changeover and machine set-up received special attention. The changeover and set-up processes were studied using critical path analyses and a cascade diagram. New tools and instruments were purchased and held in trolleys at each line so that no technician would ever be delayed because he was waiting for a tool or instrument being used in another place. Nuts on machines were replaced by new components that would not vibrate loose, causing the machine to drift out of adjustment.

For each line, the manager received a daily report that told him how many products were present as work in progress and how many products had been shipped to the distribution center. By transposing the terms in Little's Law, he produced the following formula:

$$Cycle\ Time = \frac{Work\ In\ Progress}{Throughput\ Volume}$$

His reasoning was that as he and his team worked on the production lines, their efforts should reduce variation and, therefore, cycle time. He already had daily measures of work in progress and throughput volume, so calculating cycle time daily was simple. He then plotted his daily cycle time figures. When the plot drifted down, it told him that the work done to improve the process was working. When the plot of cycle times did not fall, he knew that something was missing from their efforts.

In only a few months, he nearly doubled the capacity of some lines. Aisles in the plant that once were cluttered with trolleys of work in progress were clear. Changeover time and machine set-up time plummeted, as did downtime. Quarantined batches became a thing of the past and deviations dropped to record low levels.

Good quality, high productivity, low unit costs, and excellent customer service levels; we can have it all, if only we know how to reduce variation.

ONE POINT LEARNING

1. There is no nexus between quality and quantity. We can have it all if we know how to reduce variation.
2. Reducing variability in the flow of work increases output.
3. Six Sigma demands reduced variation, for very good financial reasons.

REFERENCES

1. W.J. Hopp and M.L. Spearman, *Factory Physics*, McGraw-Hill, New York, 1996.
2. J. McConnell, *Metamorphosis*, Quantum House Limited, Brisbane, 1997.
3. Six Sigma Seminar, sponsored by Motorola University, M. Harry et al., Sydney, 1991.

9 Case Studies

From the Pharmaceutical Industry

In this chapter, we hope to provide a guide for readers as well as some examples of what has already been achieved in the industry.

BIOLOGICAL CASE STUDY — FERMENTATION

INTRODUCTION

The protein that would be enzymatically cleaved to form the therapeutic protein was grown in a bacterial fermentation process. At the time this project commenced, most of the focus was on the more technically demanding events in the process. In accordance with the lessons from Little's Law, it was decided to shift the focus to fermentation, the first event in the process. The aim was to provide better uniformity of both quality and quantity to the remainder of the events.

Figure 9.1 shows two of the key variables, specific potency and pH, before the project was launched. Both displayed levels of variation that were higher than considered optimal.

APPROACH

The approach chosen for this project was to conduct a steady-state trial, the aim of which was to drive the process into a state of control with minimum variation. This required a shift of focus away from the location of the mean for the key characteristics onto the degree of variation. A project team was created to conduct the trial. It consisted of the production manager, a statistician, two supervisors, and four senior operators, as well as the supporting laboratory manager and an analyst.

The team started with the idea of creating as close to exact repeatability as possible, batch after batch. Some of their major activities were as follows:

1. **Raw materials.** For the period of the trial, only one batch of each raw material was to be used. Where this was not possible, Purchasing would alert the team to any batch changes before they occurred. In practice, some batch changes did occur during the trial, but their impact on the process was negligible.
2. **Seeding.** Only one bacterial working cell bank would be used to seed the fermenters for the period of the trial.

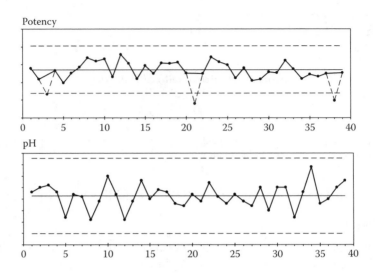

FIGURE 9.1 Pre-trial results — potency and pH.

3. **Fermenter set up.** Some differences between operators had been detected during planning for the trial. This issue was given high priority from the outset. Each step in the fermentation process was appointed a method master, usually a senior operator, who was responsible for ensuring that identical procedures were followed across all shifts. Training was conducted to ensure all supervisors and operators were clear on the aim of the trial as well as their duties. This aspect of preparation for the trial was so exhaustive and comprehensive it took two months to complete. The production manager noted that the senior operators appointed as method masters were much more demanding in terms of creating uniform operating procedures among the workforce than were the department supervisors. This was an important lesson. Initially, the production manager was surprised at how zealous and determined the methodmasters were in driving out any form of operator-to-operator variation, but concluded that they had only demonstrated how underutilized they had been in the past.

4. **Automatic process controllers.** These controllers would be set to follow the same profile, batch after batch. No adjustments would be made based on recent results. It was suspected that the controllers were making uncalled for adjustments, and it was necessary to test this hypothesis.

It is noteworthy that most activities were focused on inputs and initial set up of the fermenters. This is normal for a steady-state trial. In addition, the team was less concerned about the mean for their quality and process characteristics than they were about the variability in the process and the product.

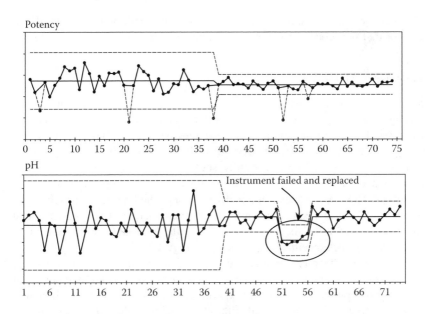

FIGURE 9.2 Results of steady-state trial — potency and pH.

RESULTS

Figure 9.2 illustrates the dramatic reduction in variation in both characteristics. Potency shows two special causes. These were identified quickly and corrective action was taken. In one case, the special cause would have been missed in the pre-steady–state trail period. The pH chart shows a significant drop in variation, as well as a period where a control instrument failed. Once this instrument was replaced, the pH returned to its stable state. This would have been more difficult to identify had it occurred before the trial, given the higher level of variation prior to the project initiation.

The steady-state trial demonstrated that the major sources of variation before the trial were operating procedures and the automatic process controllers that were over-controlling. Other areas that were improved included analytical error and sampling techniques.

The follow on effect to the downstream processing strips included higher yields and improved purity. As one operator noted, it was hardly necessary to measure the broth to notice the improvement. He could see and smell the variability in the broth before the trial. During the trial, it was so uniform that it was impossible for him to detect any difference between batches.

PARENTERALS OPERATION CASE STUDY

INTRODUCTION

A vial-filling operation was experiencing failures caused by creasing of the metal rim of the cap when it was fastened to the vial. Usually, this creasing did not cause leaks, but the creasing was deemed an unacceptable risk and was considered an indicator that the process was not performing as it should.

CREASING OF METAL CAPS

The technical people at this operation were concentrating their efforts on eradicating creasing at the capping machine itself. After much work and limited success, the focus was changed to the components. Samples were taken and all were found to meet specifications. Statistical analysis of vials and caps were undertaken. The rim of the neck on the vials was carefully studied and it appeared possible that the distribution of this characteristic was multi-modal, although the data were not conclusive. The vial neck measurements were separated based on which cavity of the relevant die had been used during their manufacture. The data immediately demonstrated that of the six cavities, one was quite different from the others despite the fact that all met specification, as noted in Figure 9.3.

What had not been understood is that faster, more automated machines tend to be less tolerant of variation in inputs than are slower or manual methods. If older machinery is replaced or supplemented with equipment that is faster, automated to a greater degree, and close coupled with other machines, the variation of the inputs must nearly always be reduced to maintain quality and quantity of output. This kind of issue is quite common in the pharmaceutical industry, particularly in parenterals and medical device operations.

In this case, the supplier was asked to capture the vials from cavity 3 and keep them separate from the remainder. The vials from cavity 3 were sent to the older line only. Once they had been removed from the new process, the creasing problems all but disappeared. The problem had been solved, but at a cost. The supplier

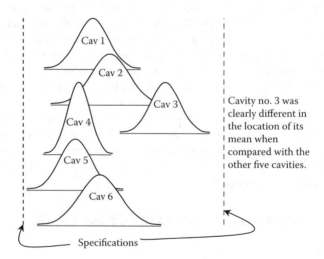

FIGURE 9.3 Distributions for vial neck rims by die cavity.

was meeting specifications, and was entitled to charge the parenterals operation for the extra work required to capture and separately ship the vials from cavity 3. This problem could have been designed out during the supplier selection process for the new line, but no one thought to test the vials for uniformity. Nor was it believed to be necessary because the vials were already in service and meeting specifications.

That type of thinking will no longer sustain businesses. As production lines become faster and more automated, it will be necessary to choose suppliers who not only meet specifications, but also who are able to ship product that is stable with Six Sigma levels of variation.

CLOSE-COUPLED MACHINES

In parenterals and medical device production and assembly lines, it is common to find a series of machines that perform different operations coupled closely together. Sometimes, the work in progress between machines is so small the transfer time is measured in seconds. In these circumstances, an upset in one machine can shut down the entire line in under a minute. This circumstance requires negligible reject rates, extraordinarily robust equipment, uniform inputs, and extremely high first-pass yields.

In the pharmaceutical industry, the practice of close coupling machines in a process should be challenged. Upsets often lead to regulatory issues as well as lost production. A superior approach may be to build in several minutes work in progress between machines, so that a small upset that can be cleared quickly does not stop the entire line. This is not a tolerance of upsets, the only acceptable level of which is zero. It is recognition that we do not live in a perfect world. Where possible, issues such as this should be designed out of the process.

Many similar processes, such as lines that bottle wines or sodas, have a couple of minutes of work in progress built in between steps in the process for similar reasons. People in parenterals and similar operations could learn much from such businesses, but it is common to find people benchmarking or studying the pharmaceutical industry only. This can be a mistake. It can be instructive to discover how different industries deal with similar issues. If best practice for bottling is examined in the wine industry, an operation is found where every wine type uses the same shape of bottle and labels for every product. Every bottle comes from a single supplier, as does every label. Colors change, shapes and sizes do not. This slashes changeover time between products to almost negligible levels and drives upsets to extraordinarily low levels. Can the pharmaceutical industry learn from such operations?

In this example, the chief issue was excessive variation in the input vials. There were several other problems that required attention, but providing the machinery with uniform inputs was the single largest obstacle to higher productivity and lower costs. Its solution required satisfactory analytical techniques to isolate the problem and a mindset that replaced the old go, no-go approach with one more in tune with the original Six Sigma concept and the Taguchi Loss Function.

SAFETY CASE STUDY

INTRODUCTION

At this site, the first control chart produced indicated that there had been a rise in overall accidents and incidents. Subsequent charts for each work group were made to determine from where this rise had come. It was not known whether the rise was general across site or had been occasioned by a rise in a single work group. Figure 9.4 shows the overall accident and incident rate, as well as separate charts for three of the major work groups.

LESSONS LEARNED

The charts were clear. The special cause in the overall chart had come from Work Group Two. The root cause for this event was never determined, as it was over four years old. Because each accident or incident was treated as a special cause, the incidents for that month were not investigated for any common links between them. The rise in the overall incident rate was solely due to Work Group Three. As absurd as it sounds, no one had noticed this rise. This was a consequence of each month's data being compared to the previous month rather than looking at the overall pattern and shape. If one month had fewer accidents than the previous month, the assumption was that safety performance had improved. If the current month showed more

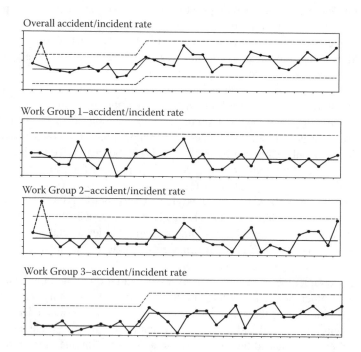

FIGURE 9.4 Safety overall and by work group.

incidents than the last month, a cause was sought, corrective action was taken, a report was filed, and incidents nearly always fell the next month.

This ought not to be a surprise. When data are stable, a characteristic significantly above average will nearly always be lower when next examined. This is not causal. It is probability theory in action.

Recall the Red Beads Experiment. During the conduct of the experiment, the manager waits until a red bead count of nine or more occurs. He then threatens the worker with dismissal if that worker does not improve on the next day. The worker almost invariably does improve. With the authors' equipment, the probability that any worker will score nine or higher on two consecutive days is only about 2 or 3%. The reduced red bead count has nothing to do with causality and everything to do with random variation and probability. Nevertheless, examples abound where similar superstitious learning has taken root and flourished.

Assume accidents, incidents, or deviations were lower this month than they were last month. The question to be asked is, "Did anything improve or are we seeing only stable, random variation?"

An ability to answer this question correctly will be just as vital to the future success of activities designed to improve safety performance as it is to production, laboratories, production scheduling and planning, and all other activities.

IMPROVED CONTROL OF POTENCY

INTRODUCTION

In this example, a drug produced offshore is imported in batches of varying sizes. It presents to the local parenterals process in crystalline form. This crystalline form is tested for potency and the dilution rate is calculated. A single batch of crystalline drug might yield two to five sub-batches of diluted product.

After dilution, the sub-batch is tested again. Once the drug is dispensed, samples are taken and final release tests are conducted. The data in this example are the potency analysis for the diluted sub-batches before dispensing. Variation was excessive and both compliance and productivity issues existed.

The diluted sub-batches were prepared sporadically. Two sub-batches per day might be prepared for two to three days, followed by a week without production, and then perhaps one or two sub-batches would be prepared on a single day. Several more days might pass before another sub-batch was diluted.

For convenience, each of these groups of sub-batches that were produced consecutively from a single batch of crystals will be called a "run." Sometimes a run will consist of a single sub-batch. At other times, a run could consist of up to four or five sub-batches.

INITIAL ANALYSIS

The first control charts made of the sub-batch test data can be seen in Figure 9.5. It is unstable, almost to the point of exhibiting chaos. Two charts were prepared.

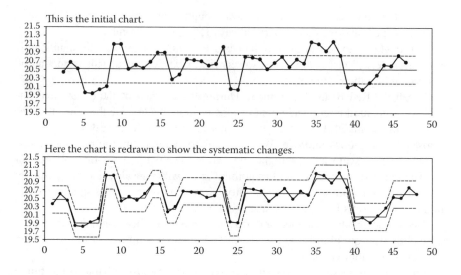

FIGURE 9.5 Potency at parenterals — dilution process.

The first chart treats the process as a single population. The second has been redrawn to display the shifts in process mean.

Nearly all of the changes to the process mean coincide with the beginning or end of a run. With one exception, every new run caused a shift in the process average. In any given run, the batch-to-batch variation was acceptable. However, the variation between runs was excessive, sometimes causing compliance issues.

The run-to-run variation implicated the dilution process. When interviewed, the plant chemists claimed that high analytical error was chief among their problems, but the laboratory control data did not support this notion. If the variation had been from batch to batch, then shift-to-shift variation, process set-up, or analytical error would have been likely culprits, but this was not the case. The subsequent investigation checked the formulae being used to calculate the dilutions, and they were found to be in order. The fact that the calculated potency did not correlate well with the actual diluted potency led to a more detailed investigation into the sampling and analysis of the crystalline compound.

The crystalline drug was visually inspected. Color variations that resembled the layers of color in pale yellow sandstone were evident. If this variation could be detected visually, the laboratory results would likely detect it as well. They did. The core of the problem was that each batch of the crystalline form of the drug was not homogenous. The samples taken of each imported batch were detecting this variation, or lack of homogeneity. The process chemists had assumed that the test results were representative of the batch of crystalline product. They were not. The dilution process improved the homogenization of the product, resulting in the poor correlation between the test results for crystals and those for the diluted product.

In summary, two issues existed. First, the crystalline batches were not homogeneous, resulting in significant sampling error. The dilution rate was being adjusted

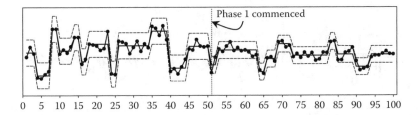

FIGURE 9.6 Sub-batch potency before and after changes to homogeneity and sampling.

based on this error, leading to the shifts in process mean noted in Figure 9.5. Second, even laboratory controls vary to some degree due to test error, and this analytical error was driving variation in the dilution rate.

ADDRESSING THE PROBLEMS

Phase 1 of Improvements

The first step was to request the offshore manufacturers of the crystalline compound either to conquer the variation in their production process or to homogenize the product before shipment. Soon afterward, improved sampling procedures were implemented at the local operation to make the samples more representative of each batch. In addition, more samples were taken to better characterize the batch. The reduction in variation was significant. The chart in Figure 9.6 shows the potency of the diluted sub-batches before and after these changes.

Phase 2 of Improvements

By necessity, the plant chemists were reacting to sampling and analytical error in the test data when conducting their dilution calculations. Once these errors are embedded in the test results, they can never be filtered out. When excessive sampling and test error is present, this additional variation will drive unnecessary or inaccurate changes to process conditions. In turn, this can lead to over- or under-control and a further increase in variation.

The laboratory manager successfully addressed analyst-to-analyst and instrument-to-instrument variation. He placed a particularly heavy emphasis on making the sample preparation process uniform across all shifts and all analysts. Chapter 17 will address laboratory issues in more detail.

The control charts for laboratory controls indicated a reduction in analytical error. The laboratory controls entered a reasonably stable state with lowered overall variation. This reduced the total variation in the analytical results for the crystals, and made the dilution calculations more reliable. Variation in the diluted sub-batches dropped yet again, as noted in Figure 9.7.

In essence, the regular shifts in the process average were being driven by a lack of homogeneity in the crystalline compound, and by inadequate sampling of this

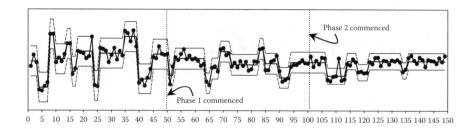

FIGURE 9.7 Sub-batch potency after second improvement phase.

material. In addition, analytical error was detected and reduced, but it played a less significant role in creating the problems encountered. Remember, the laboratory never tests the product or the process; it tests the sample. It is common to discover that a large proportion of measurement problems can be traced to sampling and sample preparation.

It is necessary to have estimates of the degree of variation caused by each of these elements. This knowledge, along with an ability to reach Six Sigma levels of variation, is vital for any business wanting to become a world-class pharmaceutical manufacturer.

DEVIATIONS IN A PHARMACEUTICAL PLANT

Deviations for microbe count on certain surfaces were triggering compliance investigations on a regular basis. The technical people were buried in deviations and doing little else. The investigating consultant examined the data for microbe counts and noted that the specification limit seemed very tight. He then interviewed the biologists who set the specifications during validation. The specification limit had been set at only 1.5 σ above the average. The biologists validating the plant intentionally did this. Their thinking was that such a tight limit would apply pressure to achieve technical excellence, as well as provide early warning of trouble. In practice, all this tight limit did was to trigger many unnecessary deviations that consumed the time of technical people who might otherwise have been involved in improving the process.

A superior approach would have been to calculate 3σ-limits and to add a further buffer to allow for the inevitable rise in microbe counts once the plant became fully operational. Finally, it would be necessary to test these new limits for safety and to have them, and the logic used to create them, cleared by the relevant authorities.

No doubt the validation biologists were well intentioned, but good intentions are the stuff with which the road to hell is paved. There is no substitute for sound science.

10 The Camera Always Lies

Knowing What Data Reveals or Conceals

Statistics is a much maligned and mistrusted science. For as long as we have had data, so too have we had manipulation and misunderstanding. Nonetheless, understanding data is an essential skill for managers.

The budding amateur wandered over to where the photographer was packing away his equipment. A conversation between the two struck up and soon the amateur was quizzing the man whose skills he hoped to some day obtain. The photographer locked his last case, lit a cigarette, ran his hands over the stubble on his face as he turned and offered some advice to the beginner. "There are only a few essentials you need to remember, the rest is detail you can pick up along the way. First, all a photographer ever does is play with light. Second, there is no such thing as correct exposure, focus, or composition. Finally, the camera always lies. The trick is to make it tell the right kind of lies. When you understand all that, when you know why it is true, and when you can control all aspects of the image, you'll be a photographer."

Anyone who has ever wondered where the warm golden glow in their late afternoon photographs came from or wondered why a face that appeared well lit to the eye shows deep shadows in a carefully exposed image knows this professional spoke the truth. The camera always lies. That's how fashion and glamour photographers make a living. Nevertheless, even an image made without any artificial light or filters and which is not retouched in any way is never exactly the same as what the eye saw through the viewfinder. It can't be because the eye and the brain are not machines and the camera is not a living organism. Natural light at noon is white light. Artificial indoor lighting produces an orange or greenish cast. Usually this color cast is not noticed because our eye and brain automatically correct the color balance to give a "normal" image. One is forced to wonder if seeing really is believing. A good photographer can produce images that look natural, if that is the desired result, but exact duplication is simply not possible.

When a photographer makes a portrait, several decisions must be made. Is the background important? Should it be kept in focus or should it be deliberately blurred to highlight the subject? If the subject has a round face, lighting from the side will make the face look longer; if a long face is encountered, frontal lighting will make it look fuller. A soft focus filter will smooth out slight facial blemishes. A light behind the subject will create highlights in the hair and help separate the subject from the background. None of this is relevant when the photographer wishes to produce a landscape.

In a similar manner, people in business must decide what information they need and then gather the data that will provide this information. There are many ways of measuring anything. Each measurement approach reveals some things and conceals others. It is not so much a question of which data are accurate or precise because no data are wholly accurate or without precision error. First, it is necessary to answer Scherkenbach's questions: What do we want to know? Why? What will we do with the answers?[1] Once these questions are answered, it is possible to then examine the potential approaches to collecting and analyzing data.

All businesses collect data and information. When people measure something and record the outcome, they are doing a job similar to that done by the photographer and the camera. For similar reasons to those in the previous paragraphs, the data can never be exact. There are many ways to measure customer satisfaction or service. Each approach gives different numbers. Even changing from attributes data to variables data can significantly alter the information content of the data. It is possible to understand what information is in the data and, just as importantly, to understand what information is not available from the data. Unfortunately, exact or complete measurements that reveal all there is to know are a dream.

The camera always lies, yet many managers can be found living most of their work lives cut off from the core work of the business except for regular reports. The data in these reports can never give a precise understanding of what is happening in the workplace. No data are comprehensive. To understand fully what is happening or what is in need of attention, managers should go to the workplace, to where the action is, in order to find out what they need to do to create improvement.

Some things are best learned through personal experience, through the physical senses rather than at an abstract level. Management literature is littered with examples where leaders have gained valuable new insights into why their businesses are struggling when they have been required to man the trenches themselves. In some companies, all managers are required to work alongside front line operators periodically. This is done to ensure that leaders have a first-hand understanding of the work and the difficulties operators and supervisors experience.

Good numerical data are essential in any business, but they ought to be supplemented by the type of data gathered by the senses; data gathered by being where the action is. In a bygone era, military officers would have called this "marching to the sound of the guns." However, even this or similar methods such as Management By Walking Around will fail more often than not if managers do not know where to look or what questions to ask.

IN GOD WE TRUST...

All others must have data. This is a popular axiom among those who rely on statistical analysis. A scientific approach based on statistics needs good data, but it is as well to remember that some things defy proper measurement. How do we measure pride and joy in work? How will we measure the benefits of education and training, or the value of customer delight and loyalty? We know these things are important, and it is possible to take some rough measures that might correlate with these aspects, but measurement with reasonable precision is not possible. If we pursue these things, we do so because they are the right thing to do, according to our values and beliefs.

HOW EXACT IS EXACT?

Experiments like that with the red beads are so controlled that we tend to assume that data from such strictly controlled environments must be fairly exact and reliable. Sadly, this is not the case. Deming, who made the Red Beads Experiment famous, used a mix of 20% red beads and a paddle that took a sample of 50. During the debrief on the exercise, he would display the data, explain the actual count of red and white beads, and ask the audience: "If we were to repeat the exercise many times, would we expect the average to settle down to some number?" After a few of us had made fools of ourselves by calculating the expected average, 10 per paddle or 20%, Deming would answer his own question, "Yes, it will settle down to some number because the process is stable. If it were unstable, the average would not settle down to any number, but would be erratic." He would then deliver the second question, "In the example with the beads, what might that average be?" Unfortunately, many of us in the audience did not detect the gleam in his eye or the beginnings of an impish grin. Again, some would shout "10 per paddle" or "20%." The elderly genius would shake his head, and seek more responses. Usually, he had to answer his own question. "It is unknown and unknowable until the trials are done and the data are collected and analyzed." Our minds raced. What did he mean? We knew that his equipment used 3200 white beads and 800 red beads (or 20% red). We knew that he took a sample of 50 each time. Why was it not possible to predict the long-term average?

Deming would then explain that his first sampling paddle was made of wood; that he had used it for about 30 years, and that it had settled down to an average of 11.3 red beads per paddle or about 22.6% red. The paddle he used through the 1980s and into the 1990s until his death was constructed of a tough synthetic material. It settled down to an average of 9.4 per paddle, or about 18.8%. What had changed? The beads and method were unaltered, but the paddle was different. Who would have thought that the beads would interact with a synthetic paddle differently than they would with a wooden one? Two days after his first Deming Seminar, John McConnell took 1000 samples from his new red beads kit. One in six of his beads were red, or 16.7%, and 5.3 per paddle. Yet, by the end of the day, the long-term average settled down to 5.7 per paddle or about 17.8% red. Twenty years later, the average remains unaltered.[2,3]

Deming taught these lessons to many thousands. Mechanical sampling, such as in the Red Beads Experiment, will provide different results if any part of the mechanism is changed. In Deming's case, the red beads were slightly larger and heavier, by an amount roughly equivalent to the extra coating of pigment the red beads received. In McConnell's case, the white beads were not as smooth as the red beads. What the actual causes are for the differences is immaterial. The important understanding is that there is no such thing as exact in our world. Heisenberg's uncertainty principle alone guarantees this.[4] No matter what data we collect, the sampling and measurement systems embed themselves into the data; they can never be "filtered out" at a later date.

In business, people face the same situation as does the photographer. We can lower or raise the light intensity; use special filters; open the aperture to blur the background and focus the viewer on the subject; manipulate the image during processing; or do our very best to create an image that is as close to the original as possible. Always we must bear in mind that our data are never exact. They cannot be.

Usually, it is possible to get a feel for how much sampling and analytical error exist, and to proceed knowing not only what the data can tell us, but also what it cannot tell us.

GIVING DATA MEANING

Data have little meaning until we know what was measured and how. If the methods of sampling or measurement are changed, so too are the resultant data. One pharmaceutical example derives from a potency assay for a certain biologic being run in two different laboratories. The results from these laboratories were unalike. Each laboratory followed the procedure as written. However, there were some notable differences. One particular difference was in a certain raw material that was not thought to be important. It turned out to be critical. The purpose of the raw material was to stabilize the biologic being analyzed. This raw material can be made in a variety of ways and it was discovered that each laboratory was using a different preparation method for the substance. This difference in preparation was the source of the bias between the laboratories.

The lesson here is clear. The word *accurate* has no meaning until the method used to obtain the result is agreed upon. Similarly, words such as *representative*, *reliable*, *relevance*, *exact*, and *pure* have no meaning until a method is agreed upon. If changing from a wooden paddle to a synthetic one changes the numerical outcomes of a very controlled process, imagine how careful business people must be in a much less controlled environment.

Another common example of this problem is the presence of two or more systems of measurement. Consider, for example, the concentration of a therapeutic protein being measured by HPLC and ELISA (immunoassay) for a fermentation sample. Both methods measure the concentration, although by different mechanisms. Both laboratory analysts claimed their methods gave good results. Both exhibited acceptable precision, but their accuracy differed.

Whose results were superior? Neither and both. What is not necessary is the unproductive argument about which set of figures is correct. Precision is the important criterion. The method that is ultimately selected should be the one providing the best precision (currently demonstrated or potential).

Where more than one shift is operating, it is common for each shift to conduct tests and measurements differently. Usually, when alignment of all shifts is attempted, much debate about which approach is superior or more accurate follows. For management, the challenge is to make very clear that it cares little for "better" or "accurate" at the outset. What is required at the outset is uniformity or repeatability. Having everyone using the same approach and obtaining comparable results is nearly always a first step. Once this has been achieved, it is easier to conduct trials to determine which methodology is superior.

SERVICE INDUSTRIES

A manager who had been appointed head of a distribution business noted that the customer service levels of different centers varied widely. Initially he focused his attention on those centers with the poorest results in the monthly reports. After a while, he discovered that there was very little real difference between the centers. Some managers pulled their data out of focus by moving certain outcomes from one month to another. Others used filters to smooth out undesirable blemishes.

At a conference, this executive had managers from each center flowchart their order entry and dispatch process. No two were alike. They were not even measuring the same thing. The data could not be compared because they were results not only of different processes, but also of different methods of measurement.

An interesting example comes from the telecommunication industry. Measures of customer complaints and survey data to establish customer satisfaction levels were available to management. Again, there was a significant difference between regions. In this case, careful analysis revealed that a large part of these differences had less to do with actual performance than it did with customer expectations. Regions servicing relatively wealthy suburbs showed different customer satisfaction levels than those operating in less affluent suburbs, even when the operational performance data seemed comparable. In hindsight, the managers realized that the differing expectations of people from varying socio-economic backgrounds by their nature must alter the data. As the people conducting the surveys switched their attention from one suburb to another, they were measuring different systems. The data were not comparable because they could not be.

Survey data gathered from verbal or written responses from customers or employees are particularly vulnerable, and should only be gathered and analyzed by people well schooled in the subject. Humans are emotional, subjective, intuitive creatures. We are seldom entirely cool, rational, and objective. Our responses to surveys reflect our emotional state at the time.

The temptation to rank measurements from different sites, laboratories, or factories is one that ought to be resisted. This is particularly the case with service industries because there is so much room for differences in the measurement systems. Part of the differences will always be in the measurement systems themselves. Of more importance is the determination of whether the data are stable and whether they are trending. In addition, when data from different operations are routinely rank ordered, there is the temptation among those being ranked to use a rose-colored filter.

Even establishing trends is done poorly in many companies.[1,3] In many businesses, a cause is sought if this month's figure is higher or lower than it was last month, even when the data exhibit stability.

As the old axiom states, data is like an Italian swimsuit. What it reveals is interesting. What it conceals is vital. In both cases, only the observer who has at least a general understanding of what is concealed can fully appreciate that which is revealed. Like Italian beachwear, good data are valuable and expensive, but even good data are limited in their information content.[1-3]

We tend to be too deterministic in our use of data, and would do well to remember the professional photographer's advice to the novice. When was the last time you checked your camera?

ONE POINT LEARNING

1. There is no such thing as an absolute or true value in our world.
2. No data have meaning until the sampling and measurement methods have been operationally defined.
3. Knowing what information cannot be extracted from the data is as important as knowing what information can be gleaned from it.

REFERENCES

1. W. Scherkenbach, *Deming's Road to Continual Improvement*, SPC Press, Knoxville, TN, 1991.
2. W.E. Deming, *Out of the Crisis*, MIT Press, Cambridge, MA, 1988.
3. J. McConnell, *Metamorphosis*, Quantum House Limited, Brisbane, 1997.
4. S. Hawking, *A Brief History of Time*, Bantam Books, New York, 1988.

11 Keeping It Simple

Finding Pattern and Shape in the Data

At the level at which this book is pitched, the initial statistical analyses undertaken need not be difficult. Indeed, they may not involve anything more sophisticated than a plot of points and a clear plastic rule.

TIME — THE FIRST IMPERATIVE

A control chart always plots data in time sequence.[1] Shewhart noted that a significant amount of the information sought in industrial applications is contained in the relationship of the data to each other when they are plotted in time order. Many types of statistical analyses, such as frequency distributions and simple two-variable correlations, do not show the data in time order. Such statistical tools are still valuable, but the information contained in time sequence is essential.[2] Figure 11.1 shows two charts for a single set of data, a histogram or frequency distribution and a time series plot. The time series plot reveals a downward trend that was concealed in the frequency distribution.

PATTERN AND SHAPE

In his excellent book, *Facts from Figures*, M.J. Moroney thought that finding pattern and shape in data was important enough to dedicate an entire chapter to the subject.[3] Experience demonstrates that he was correct to place an emphasis on this subject early in his book. Some things need no calculations because they are obvious when pattern and shape are examined.

Figure 11.2 shows a plot of points for the weight of pharmaceutical product extracted by a centrifuge, batch by batch. If examined closely, it reveals an almost perfect saw-tooth pattern. But what is driving this pattern? The regular up and down adjustments to a process found during over-control can cause such a pattern. If each point represented one shift, it could be shift-to-shift variation, where day and night shifts managed the process differently. Equally, it could be two machines operating at different levels. In this case, it was the latter. Where two or more machines or instruments exist, it is safe to start with the assumption that they are different. It is important to know how different, and in what characteristics. Six Sigma levels of performance will be difficult to achieve if these differences are significant. When the data were stratified and plotted as separate charts for each centrifuge, the pattern disappeared, as noted in Figure 11.2. The saw-tooth pattern

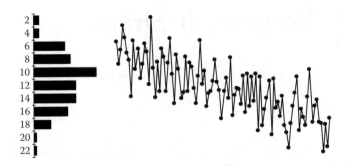

FIGURE 11.1 Time ordered data reveals trends that histograms do not.

was caused by the difference in performance of the two machines. This pattern would have been invisible in a frequency distribution, and in many other statistical techniques.

No sophisticated analysis was required to find this problem, as well as the root cause. All that was required was to plot the points and study them for pattern and shape. This approach is recommended as a beginning point for every statistical analysis undertaken.

Figure 11.3 shows two time series plots for the batch run time for a process in a pharmaceutical plant where alcohol is evaporated to precipitate the product. An average is drawn on the top chart. Initially, this average can be either estimated

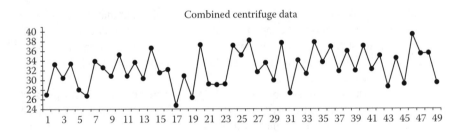

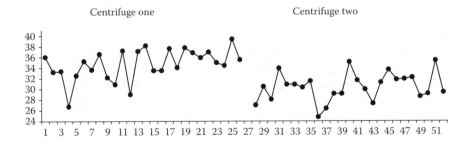

FIGURE 11.2 A saw-tooth pattern in centrifuge production data.

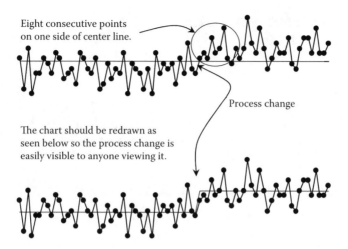

FIGURE 11.3 Process batch run time — shift in process average.

or calculated. From Chapter 5, readers will recall that for the data to be stable the points must fall at random about the center line. The key phrase is "at random." If seven or eight consecutive points fall on one side of the center line, this is deemed to be statistical evidence that the data are no longer random. Such a signal indicates that the process has changed and has shifted its mean. In this case, such a signal is evident. The process average has shifted to a higher location.

Figure 11.4 shows a plot of the number of alarms per day for a pharmaceutical production line. The two special causes are reasonably clear. Again, there was no need for calculations to determine that these two points looked different from the remainder. As will be seen in later chapters, these signals are even easier to see on a moving range chart.

During seminars and courses, it is recommended that before any calculations are performed, the early steps in data analysis should be the following:

1. **Plot** the points in time sequence. This should always be the first step.
2. **Study** the points for pattern and shape, and for any evidence of non-randomness. Look for repeatable patterns and for obvious "flyers" in the data.

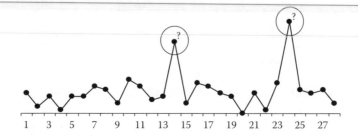

FIGURE 11.4 Two special causes evident.

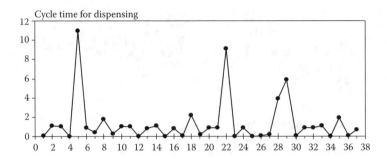

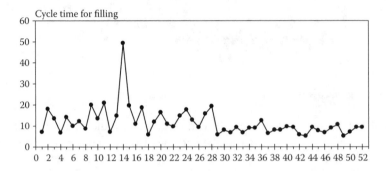

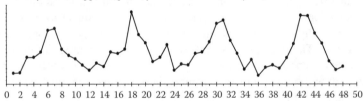

FIGURE 11.5 Example charts.

3. **Estimate** or calculate a center line and draw it on the chart. During our courses, students are asked to estimate the center line with a clear plastic rule, and many surprise themselves with their proficiency. Look again for pattern and shape. Are any other signals evident? Determine whether seven or eight points fall on one side of center line at any stage.
4. **Calculate** control limits. Then apply the tests for stability. (More tests for stability will be covered in later chapters.) Once again, study the chart for pattern and shape.

THE DTLF (DARN THAT LOOKS FUNNY) APPROACH

Issac Asimov said:

> The most exciting phrase to hear in science, the one that heralds discoveries, is not 'Eureka!' but, "Now that's funny…"[4]

In one pharmaceutical company, a young supervisor independently reached the same conclusion. Soon after learning Shewhart's methods, she realized that the phrase that most commonly preceded a breakthrough or a new understanding was, "Darn that looks funny." This kind of comment is common when people are studying data for pattern and shape. It will be absent when the old go, no-go approach rules.

Some example plots used in our courses and seminars are shown in Figure 11.5. They are presented here as an exercise for readers. Those who undertake the exercise should study the charts for pattern and shape to determine what signals might be detected without the aid of calculations or control limits. Any suspected special causes should be circled and center lines drawn on the charts to indicate the number of populations present. The charts in Figure 11.6 show the interpretations of the authors.

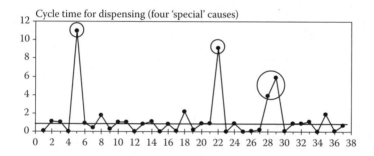

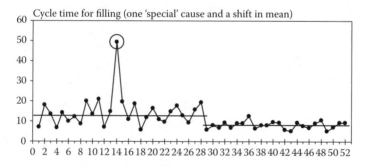

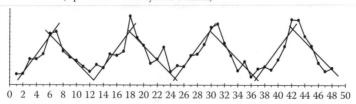

FIGURE 11.6 Interpretations of example charts.

REFERENCES

1. W.A. Shewhart, *Economic Control of Quality of Manufactured Product*, Van Nostrand, New York, 1931.
2. W. Scherkenbach, *Deming's Road to Continual Improvement*, SPC Press, Knoxville, TN, 1991.
3. M.J. Moroney, *Facts from Figures*, Penguin Books, London, 1951.
4. G.E.P. Box, J.S. Hunter, and W.G. Hunter, *Statistics for Experimenters: Design, Innovation and Discovery*, 2nd ed., John Wiley & Sons, Hoboken, NJ, 2005.

12 Why Use Control Charts?

An Introduction to the Types of Charts

The Shewhart control charts largely overcame four of the shortcomings in existing statistical methods. These, along with some advantages for using control charts, are discussed here.

WHY USE CONTROL CHARTS?

Why would someone want to use a control chart? Essentially, a control chart is used to separate random from non-random variation; to indicate when variation is stable and when it is not. This leads to three interlinked approaches:

1. To analyze data in order to understand it and the process involved
2. To aid in attempts to reduce variation (to indicate what course of action is called for and which attempted improvements worked and did not work)
3. To control the process on a minute-by-minute or batch-by-batch basis

TYPES OF DATA

For control chart purposes, data can be divided into variables and attributes. In essence, they can be described as follows:

1. **Variables.** Things that are measured. Some examples are time, weight, purity, potency, particle size, and pH. Variables operate on a continuum. Generally, but not always, the degree of precision it is possible to attain is limited only by the precision of the measurement methods. Discrete variables, those that must always be expressed as whole numbers, do exist but the difference between discrete and continuous variables is irrelevant for control chart purposes.
2. **Attributes.** These are count and classify data where discrete outcomes are classified as good or bad, in specification or out of specification, on time or not on time, etc. The key word is *classification*. If the data are not classified, the data is treated as a variable.

For variables data, the charts in common use are the average and range control chart and the individual point and moving range control chart. Both are discussed in later chapters. There are four types of attributes charts in common use. Because the bulk of the data found in the pharmaceutical industry are variables data, the attributes charts will not be covered here. However, attributes data can be control charted by substituting the individual point and moving range chart for the applicable attributes chart. It is likely that the control limits will be spread a little wider than would be the case if the correct attributes chart had been used, but this is not a significant disadvantage. Statisticians and some people in the QA function already understand attributes charts, allowing them to be applied where necessary.

ADVANTAGES OF CONTROL CHARTS

During his early studies, Shewhart noted that the existing statistical tools had certain shortcomings, four of which are listed below. Shewhart's control charts went a long way toward overcoming all four.

1. **Average and range are not linked for variables**. For variables data, the central tendency (location of the average) and the degree of dispersion (or range) are not necessarily related. Each characteristic can change independently of the other, as noted in Figure 12.1. It for this reason that the Shewhart control charts for variables use two charts to track each characteristic separately.[1]
2. **Shifts in process average.** Individual data are not sensitive to slight but sustained shifts in the process center line. Plots of averages of small subgroups are much more sensitive in detecting small changes in the process mean. This aspect is covered in detail in later chapters.[2]
3. **Non-normal distributions.** Many distributions are not normal. As will be seen in later chapters, sample averages always form a bell curve if

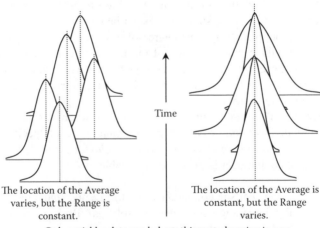

The location of the Average varies, but the Range is constant.

The location of the Average is constant, but the Range varies.

Only variables data can behave this way, changing in one characteristic, but not necessarily the other.

FIGURE 12.1 Average and dispersion operate independently for variables data.

stable. This means that the functional form or shape of the distribution ceases to be a significant issue either in interpretation or in the development of tests for stability. Normality is not a requirement for the Shewhart control charts to work,[3] but the fact that sample averages approach normality simplifies interpretation greatly.

4. **Calculating control limits.** Calculating the standard deviation of the plotted values is a problematic approach for developing control limits. The common practice of setting limits at two or three standard deviations of the plotted values on either side of the center line can make the most chaotic data look stable. Shewhart's control charts for variables overcome this problem by developing the control limits from the ranges chart. This is explained in some detail in Chapter 13 and Chapter 14.

DEVELOPING CONTROL LIMITS

In the examples in Figure 12.2, the left-hand distribution was created artificially. For the purposes of this section, stability can be assumed. The average ($\bar{x}$) and the standard deviation (σ) were calculated. Finally, limits were calculated at $\bar{x} \pm 3\,\sigma$.

Some special causes, shown as black discs, were added to the distribution seen in the right-hand side of Figure 12.2. The calculations were repeated as per the left-hand distribution. In both cases, the calculated limits captured all the data. If even more special causes were added, it is likely the limits would spread further apart to capture these data as well.

This happens because of the way standard deviation (or root mean square deviation) is calculated. First, the deviations are calculated by determining the difference between each data point and the mean. These deviations are then squared. The mean or average of these squared deviations is determined and, finally, the

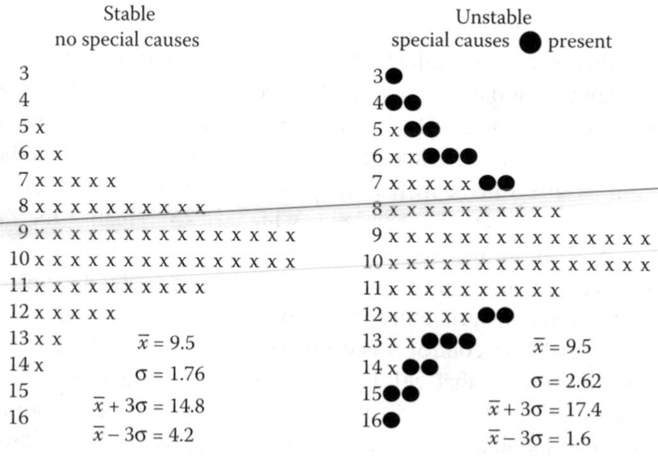

FIGURE 12.2 Developing limits at $\pm 3\ \sigma_{RMS}$ creates an illusion of stability.

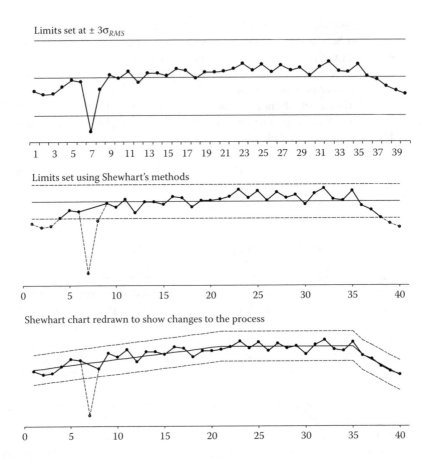

Limits set at ± 3σ_{RMS}

1 3 5 7 9 11 13 15 17 19 21 23 25 27 29 31 33 35 37 39

Limits set using Shewhart's methods

0   5   10   15   20   25   30   35   40

Shewhart chart redrawn to show changes to the process

0   5   10   15   20   25   30   35   40

FIGURE 12.3 Developing limits at ± 3 σ_{RMS} creates the illusion of stability.

square root of this mean is calculated. Because special causes tend to happen near or past the extremities of the distribution, their deviation values will be large. These large values inflate the calculation for σ, which is precisely what should happen because σ is a measure of the tendency of the data to disperse around the mean of the data. A similar effect occurs when the process mean drifts up or down.

Unfortunately, this very characteristic, which is so valuable for distribution theory, makes σ_{RMS} (root mean square deviation of the plotted values) unworkable as a means of calculating control limits. Control limits should help to separate random from non-random variation. A better way to estimate σ that is less affected by special causes and over-control is needed. Shewhart's control charts do just that, with one important proviso that will be discussed in later chapters.

This phenomenon is illustrated in Figure 12.3, which shows three charts. All charts use the same laboratory control data. The top chart has limits developed by using ± 3 σ_{RMS} and, with the exception of one point, looked stable to the laboratory staff. The lower charts use Shewhart's methods (ranges charts are omitted for clarity and simplicity). They reveal the true situation; the process is unstable. The data is drifting up and down. Given that this is laboratory control data, reason for real concern exists.

In summary, provided the distribution is reasonably continuous (no flyers or significant breaks in the tails of the distribution), 3 σ_{RMS} limits will always create the illusion of stability, even though the data are in a state of chaos.

What is clear from the charts in Figure 12.3 is that Shewhart's methods demonstrated the drift upward early in the chart, and the drift back down late in the chart. The only signal exposed by the top chart is the special cause that readers would have noted without any limits at all.

Having demonstrated how not to develop limits, the next chapter introduces the first of Shewhart's charts.

ONE POINT LEARNING

1. For variables, central tendency and variation are not connected.
2. Limits based on 3 σ_{RMS} will make any continuous distribution look stable, even if a state of chaos exists.

REFERENCES

1. E.L. Grant and R.S. Leavenworth, *Statistical Quality Control*, McGraw-Hill, New York, 1980.
2. J.M. Juran, F.M. Gryna and R.S. Bingham, *Quality Control Handbook*, McGraw-Hill, New York, 1974.
3. D.J. Wheeler, *Advanced Topics in Statistical Process Control*, SPC Press, Knoxville, TN, 1995.

13 Average and Range Control Charts

An Introduction to Control Chart Practice

In this chapter, average and range control charts will be introduced by guiding the reader through the mechanics of constructing the charts. Why and how the charts work is discussed, as is sub-group integrity.

First, it is necessary to establish what an average and range control chart looks like. An example appears in Figure 13.1. It is a chart of data from a medical devices plant assembling automatic injectors. Alignment problems had been experienced in the plant's high-speed assembly machines, leading to variable pressure required to activate the device and interruptions to the process to adjust the alignment and assembly of the components. This, in turn, led to reduced production capacity. The variable under examination is the alignment of a critical component.

The first thing to note is that there are two charts for variables. The upper chart is for the averages of small sub-groups and the lower chart is for the ranges of these sub-groups. Chapter 12 explained how variables could alter the location of their mean without changing their range, and vice versa. Because these two characteristics are not necessarily connected, they can change independently of each other, and frequently do. We need two charts so we are able to track central tendency (averages) and variability (ranges) separately.

CONSTRUCTING AN AVERAGE AND RANGE CONTROL CHART

The data are gathered usually in small sub-groups. For average and range control charts, the most common sub-group size is four, for reasons that will become apparent in Chapter 14. In this case, the existing methodology was to take three samples periodically from the assembly process, so these historical data were used at the outset. An important feature of control charts is that small sub-groups, taken at regular intervals, do a superior job of helping people to understand the process than do larger samples taken less frequently. This approach provides the information contained in the time-ordered sequence of the data discussed in Chapter 11. Table 13.1 shows the first eight sub-groups (of three data each) for the charts in Figure 13.1.

The next step is to calculate the average ($\bar{x}$) of each sub-group. This is written above the sub-group in the appropriately marked row. For the first sub-group, $\bar{x}$ = 10.41.

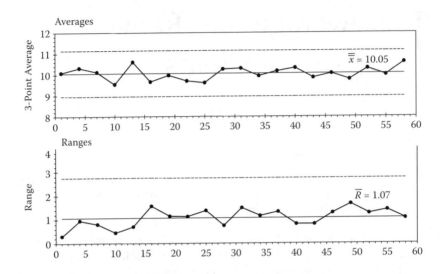

FIGURE 13.1 Average and range control chart for alignment.

Then the range (*R*), or the difference between the smallest and largest data, is calculated and written immediately under the average. The first sub-group has an *R* of 1.53. This is done for all sub-groups. This provides the two statistics of interest, a measure of central tendency (average) and a measure of variability (range). These are plotted in Figure 13.1.

Chapter 11 recommended a study of the plotted points for pattern and shape before doing any calculations for control limits. In this case, there were no obvious signals, and the next step, to draw an average on each chart, is taken. To calculate the average of the averages ($\bar{\bar{x}}$), add the averages and divide by the number of sub-groups. Repeat this process with the ranges to calculate the average of the ranges

TABLE 13.1
First Eight Sub-Groups of Data from Figure 13.1

$\bar{x}$	10.41	10.32	10.12	9.54	10.60	9.66	9.96	9.70
R	1.53	0.95	0.80	0.45	0.70	1.56	1.14	1.12
	11.30	10.85	10.35	9.35	10.60	9.71	10.59	9.87
	9.77	9.90	9.61	9.80	10.25	10.41	9.45	10.17
	10.17	10.20	10.41	9.48	10.95	8.85	9.85	9.05

The data are assembled in sub-groups
(of three in this instance).

($\overline{R}$). Draw these center lines on the respective charts and again study the points for pattern and shape. If statistically significant shifts in the means are found, split the chart into separate populations. In this case, no such signals are evident. The next step is to calculate the upper and lower control limits (UCL and LCL). The formulae are:

Average chart: $$U/LCL = \overline{\overline{x}} \pm A_2 \overline{R}$$

Ranges chart: $$UCL = D_4 \overline{R}$$

$$LCL = D_3 \overline{R}$$

The formulae require the A_2, D_3, and D_4 factors, which are a function of sub-group size (n). The origin of the formulae and the factors are discussed in Chapter 14. The factors for n can be found in Table 13.2. In this instance, the sample size is 3. The factors for this n are circled in Table 13.2. Note that the value of the D_3 factor is zero until the sample size reaches seven, meaning that the LCL of a ranges chart will always be zero until the n is seven or greater.

For this example, the calculations are:

Averages chart $$U/LCL = \overline{\overline{x}} \pm A_2 \overline{R}$$

$$= 10.05 \pm (1.02 \times 1.07)$$

$$= 10.05 \pm 1.09$$

$$UCL = 11.14$$

$$LCL = 8.96$$

TABLE 13.2
Factors for n for $\overline{x}$ and R Control Charts

n	1	2	3	4	5	6	7
A_2	2.66	1.88	1.02	0.73	0.58	0.48	0.42
D_4	3.27	3.27	2.57	2.28	2.11	2.00	1.92
D_3	0	0	0	0	0	0	0.08

Ranges chart
$$UCL = D_4 \bar{R}$$

$$= 2.57 \times 1.07$$

$$= 2.75$$

These control limits are marked on the chart, and the chart is studied for pattern and shape. In this case, there appears to be nothing other than random variation within the control limits. The process exhibits stability. Nothing is changing. Of course, the fact that the data are stable tells nothing about the capability of the process. It could be producing stable garbage, but it is predictable within limits.

During the initial analysis, the production manager stratified his data into shifts to determine whether there were any significant differences between them. There were. Careful analysis revealed that whenever a particular operator set up the machines, precision improved. Soon afterward, this operator was promoted and given the responsibility for training people and ensuring that uniform operating procedures for this process were being used across all shifts.

Figure 13.2 shows the same process as noted earlier, but with more data added immediately after this promotion.

Note the long collection of points below the average in the ranges chart. This is evidence enough that something has changed. Note also that this change is easier to see in the ranges chart than it is in the averages chart. Once this evidence is at hand, the chart is redrawn to show these changes, as seen in Figure 13.3. This allows even someone who knows little about control chart theory and practice to interpret the chart at a glance. Data for this example are in Table 13.3.

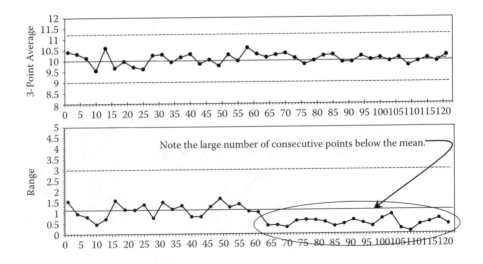

FIGURE 13.2 Further analysis of alignment.

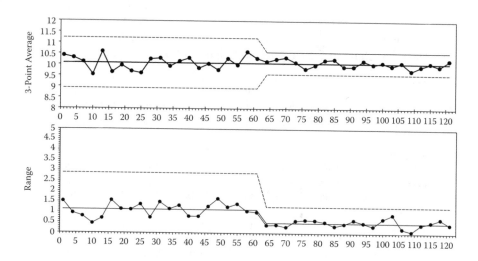

FIGURE 13.3 Table 13.2 redrawn to show the shift in $\bar{R}$.

The improvement in precision was substantial. As a direct consequence of this work, rejects fell, and volume output lifted. Uniform machine set-up resulted in an improvement in performance. It nearly always does.

HOW THE FORMULAE WORK

When the $\bar{R}$ in Figure 13.3 shifted down, the control limits for the averages chart moved closer together. This seems to make intuitive sense. The ranges chart measures variability. If variation is reduced, one would expect the control limits to move closer together, and they do. The reason why this happens is shown in the formulae for the charts.

Averages chart $$U/LCL = \bar{\bar{x}} \pm A_2 \bar{R}$$

Ranges chart $$UCL = D_4 \bar{R}$$

$\bar{\bar{x}}$ and $\bar{R}$ are the center lines for the charts. For any given population, they are constant. (A chart can have more than one population. The charts in Figure 13.3 show two populations.) D_4 and A_2 are constant for any given sample size. *Therefore, the only characteristic that can impact the spread of the control limits for any given population, on both charts, is $\bar{R}$.* This is a critical understanding that will be explained in more detail in Chapter 14.

Because all control limits are developed from $\bar{R}$, it is generally correct to say that to the extent that $\bar{R}$ is a reliable estimator of within sub-group variation, so too will the control limits be trustworthy. To the extent that $\bar{R}$ is corrupted by special

TABLE 13.3
Data for Chart in Figure 13.2 and Figure 13.3

Sub-group	1	2	3	4	5	6	7
	11.30	10.85	10.35	9.35	10.60	9.71	10.59
Data	9.77	9.90	9.61	9.80	10.25	10.41	9.45
	10.17	10.20	10.41	9.48	10.95	8.85	9.85

Sub-group	8	9	10	11	12	13	14
	9.87	10.25	9.96	9.40	10.45	10.80	10.20
Data	10.17	9.70	10.70	10.60	9.30	9.48	9.96
	9.05	8.88	10.10	10.88	10.04	10.20	10.76

Sub-group	15	16	17	18	19	20	21
	10.18	10.82	9.95	9.90	9.92	11.25	10.50
Data	9.98	9.55	8.85	11.10	10.72	10.20	9.70
	9.38	9.77	10.50	9.85	9.32	10.34	10.70

Sub-group	22	23	24	25	26	27	28
	10.30	10.04	10.50	9.80	10.20	9.80	10.33
Data	9.92	10.44	10.20	10.40	9.72	10.40	10.44
	10.24	10.32	10.32	10.16	9.56	9.78	9.90

Sub-group	29	30	31	32	33	34	35
	10.25	9.75	9.56	9.90	10.20	10.25	10.50
Data	10.10	9.84	10.20	10.40	9.85	9.70	9.80
	10.45	10.20	10.04	10.30	10.05	10.40	9.60

Sub-group	36	37	38	39	40	41	
	10.12	9.78	9.68	9.80	10.35	10.33	
Data	10.23	9.70	10.05	10.15	9.65	9.96	
	10.00	9.80	10.11	10.34	9.85	10.40	

causes or by systematic chaos in the process, so too will the position of the control limits be affected.

This understanding cannot be overemphasized. Many control charts in pharmaceutical plants are worse than useless because they mislead people; they confound the analysis and conceal more than they reveal. In many cases, the computer programs used to make the charts default to a setting where the ranges chart is not visible. If one cannot see the ranges chart and becomes satisfied that $\bar{R}$ is a good estimate for within sub-group variation, no degree of reliability can be attached to the control limits for either chart. An example of this undesirable situation can be seen in Figure 13.4.

It is bad enough when computer programs default to a setting where the ranges chart is not visible, but how much worse is it when people ignore or hide the ranges chart? This situation has been encountered in the pharmaceutical industry,

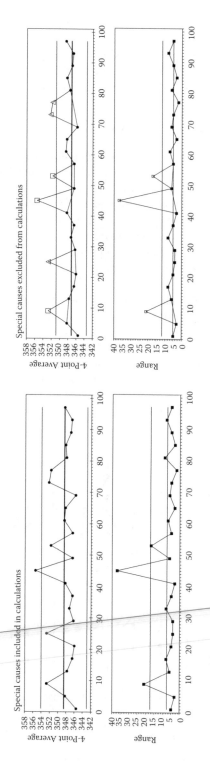

FIGURE 13.4 Illustration of the impact of special causes on x̄ and R chart control limits. In this example, an otherwise stable ranges chart was disturbed by three rogue data values. On the right hand chart, these three sub-groups of four data (each containing a single special data value) are denoted by a square point. The "spikes" in the ranges chart expose these occurrences. Also, two short-term drifts in the process mean exist, where all the data in one, and later two, sub-groups drifted up for a period. These points are indicated with a triangle. Note that such drifts in mean have no impact on the ranges chart, provided sub-group integrity is maintained (discussed later in this chapter). The increased spread of the control limits on the left hand charts is due entirely to three special causes. Broken lines connecting plotted points in the charts on the right indicate that these points have been removed from the calculations. In the pharmaceutical industry, it is common to find the averages chart without its matching ranges chart. In this case, the plant people saw the left-hand averages chart only (no ranges chart was visible on the screen or on the printed copy). They interpreted it as reasonably stable with one small special cause. Once the special causes were removed from the calculations for R, as seen on the right hand charts, a different picture emerged. The charts have 25 sub-groups. Six of them are statistically significant. These indicate unnecessarily high variation and much potential for improvement. Unfortunately, the plant manager and his staff saw only one small disturbance. If problems are invisible, they are likely to remain unaddressed, and much improvement is forsaken.

and it cannot be tolerated; not if the business aspires to Six Sigma levels of variation and world-class status.

The significance of $\bar{R}$ in establishing control limits is why the ranges chart is always interpreted first. There is no point attempting an interpretation of the averages chart before the trustworthiness of the control limits has been established.

WHY THE CHART WORKS

A control chart places control limits at ±3 standard deviations from the center line,[1-3] but it does not use the root mean square deviation (σ_{RMS}) function found in calculators or computers. As will be seen in more detail in Chapter 14, $A_2\,\bar{R}$ provides an estimate of three standard deviations of the averages ($3\,\sigma_{\bar{x}}$) based on $\bar{R}$. For control chart purposes, this is a superior approach because this method of estimating $3\,\sigma_{\bar{x}}$ will, in nearly every case, be less affected by process disturbances than is σ_{RMS}. Figure 12.2 in Chapter 12 illustrated that as disturbances are added to a population, σ_{RMS} grows and limits set at $\pm3\sigma_{RMS}$ are likely to capture all or most of the data, whether it be stable or profoundly unstable.

To illustrate how the $\bar{x}$ and R chart works, an example will be developed using data from an over-controlled process. A sample of four products is taken every hour, and the process is adjusted by an amount equal to and opposite of the error in an attempt to hit the target value. To commence this illustration of Shewhart's approach, the upper chart in Figure 13.5 uses distribution curves to demonstrate the range and location of the average for this process. Of course, we can never know the exact parameters of the process.[1,4,5] Sub-groups of data are taken to estimate these parameters, and these sub-groups are represented by the four black dots in each distribution.

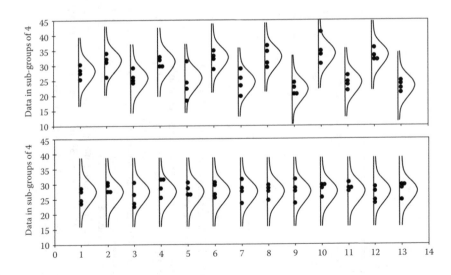

FIGURE 13.5 Distributions of 13 sub-groups where n = 4.

Thirteen sub-groups of data where $n = 4$ are shown. For comparison, the lower chart in Figure 13.5 is the same process after it was stabilized.

Note the saw-tooth pattern of over-control in the upper chart. Interestingly, the ranges, or the variation *within* each sub-group, remain unaltered over time for both charts. In this and many other cases, the variation within the sub-groups remains unchanged regardless of whether over-control is present, provided sub-group integrity is maintained. Often, the most important aspect of sub-group integrity is that the procedure followed makes it impossible for an adjustment or change to occur *within* a sub-group. Any adjustment, in this case, must occur *between* sub-groups, so sub-group integrity is maintained.

Two methods of calculating limits are displayed in Figure 13.6. In the left-hand chart, limits were calculated using $\pm 3\sigma_{RMS}$. This is a common approach in the pharmaceutical industry. The right-hand chart uses Shewhart's formulae, the only approach that can be recommended.

SUB-GROUP INTEGRITY

What has been established in the preceding pages is that Shewhart's control limits depend on the trustworthiness of our estimate for the underlying systematic process variation, or $\bar{R}$. This statistic is an average of the ranges of several sub-groups, so in turn it is dependent on the trustworthiness or integrity of each of the ranges within the sub-groups. The sub-group can be expected to retain its integrity if all the data in the sub-group are produced under essentially the same technical conditions. If a sub-group is free of disturbances or special causes, and if no change has been made to the process within that sub-group, it can be anticipated to retain its integrity. This subject is pursued in the following chapters.

Special Causes

If in a sub-group where $n = 4$ a single data value is special or different from the other three data, often it will not fall outside the limits of the averages chart but will fall beyond the UCL for the ranges chart. This situation was noted in Figure 13.4. It is important that such points be removed from the calculations for $\bar{R}$. Central to the purpose of control charts is to separate random from non-random variation. By taking special cause points out of the calculations, we improve our estimate for $\bar{R}$ and the control limits become more trustworthy.

Process Changes or Adjustments

Usually, samples are taken in such a way that changes or adjustments to the process will always fall between rather than within sub-groups. When this is done, adjustments to the process seldom have any impact on the ranges charts and sub-group integrity is maintained. This situation is generally simple to arrange with instantaneous sampling, but is problematic with serial sampling, as is discussed later in this chapter.

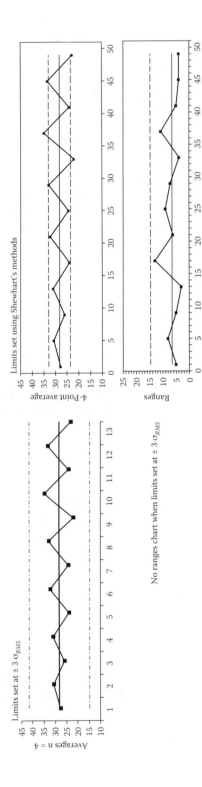

FIGURE 13.6 Over-control — plots of 13 sub-groups where n = 4. As anticipated in Chapter 12, if limits are set at ±3σ_RMS, even chaotic data are likely to remain within these limits, with the potential exception of extreme, isolated "flyers." Over-control is a common problem in the pharmaceutical industry. Given the tendency of people in the industry to use a binary "in or out" mindset, and to not carefully examine the data for pattern and shape, many would declare the upper left chart to exhibit stability. However, the Shewhart control chart on the right demonstrates instability, exposes an opportunity for improvement, and is a spur to action. In this example, the spread of the control limits was tripled by using ±3σ_RMS limits instead of Shewhart's formulas. Shewhart's method succeeded in establishing the limits of controlled variation whereas ±3σ_RMS produced limits that concealed more than they revealed. Shewhart's formulas work because they calculate the control limits from within sub-group variation (the ranges chart) rather than from the plotted points themselves. Some theory behind the formulas will be introduced in the next chapter. Strictly speaking, fourteen points must alternate up and down to indicate a lack of control (see next chapter), but in this case the ninth and tenth points fell beyond limits, demonstrating instability.[5]

Duplicate and Triplicate Sampling

The pharmaceutical industry has a plethora of situations where duplicate or triplicate samples are taken. Usually, this is done in an attempt to reduce sampling error and to gain a reasonably representative sample from batch processes. A common approach is to agitate or blend the material for a defined time interval and then to take a sample. After another period of agitation, another sample is taken, and then after more agitation a third sample is drawn. These three samples are tested and an average of the three is used as the test result for the batch. If these three samples are used to make a sub-group in an $\bar{x}$ and R chart, it is common to find the control limits so close together that most if not all points fall beyond the control limits. Essentially, the same material is being measured three times. The sampling error is present in each sub-group. Process variation is absent *within* sub-groups but is present *between* sub-groups. Because for duplicate and triplicate samples sub-group variation is exclusive of process variation, this process variation is not included in the calculations for control limits. However, process variation is present from batch to batch. The sub-group-to-sub-group variation will always be greater than allowed by limits developed using the three values from a triplicate sample as a sub-group, and such a chart will never exhibit stability, no matter how controlled the process. For triplicate samples, the proper method is to average the triplicate sample results and to plot this average on an individual point and moving range control chart as described in Chapter 15.

Instantaneous Sampling

This is the type of sampling where the entire sub-group is sampled at one time. Instantaneous sampling is used for discrete items or products. A supervisor might take a sample of, say, four cartridges or vials from a particular filling head every hour. An operator will take a production sample of, say, five medical devices from the assembly line on a periodic basis. In these cases, instantaneous sampling is being used. Importantly, it is difficult to imagine any change or upset that would affect one of these items and not the remainder. This is why, usually, maintenance of sub-group integrity is not difficult where instantaneous sampling is used, and only one process is sampled.

Serial Sampling

When a single sample is taken on a periodic basis, serial sampling is in use. The data from these samples follow one another over time, in serial. This type of sampling is more common in batch or continuous operations. It is difficult to imagine how an instantaneous sub-group could be taken in these situations. Serial data can be used to make an $\bar{x}$ and R chart, but certain fundamentals must be understood to avoid misinterpretation. One way to illustrate differences between instantaneous and serial sampling is to demonstrate how differently they can display over-control.

SERIAL SAMPLING — LOSS OF SUB-GROUP INTEGRITY AND OVER-CONTROL

Figure13.7 is a control chart for a continuous chemical process utilizing serial sampling. The flow rate is critical, and this characteristic had been problematic for several years. An automatic process controller was regulating flow rate. The decision was made to test it with an $\bar{x}$ and R control chart ($n = 4$).

If routine over-control is present in a process using serial sampling, often it is revealed by the pattern of points looking too precise. They hug the center line. The averages chart in Figure 13.7 has two sets of limits, the normal 3 $\sigma_{\bar{x}}$ limits and another set at 1 $\sigma_{\bar{x}}$, that have one third the spread of normal limits.

Every point on the averages chart falls inside the 1 $\sigma_{\bar{x}}$ limits. In many situations, there are special causes present that confound the analysis. Nevertheless, if the bulk of points fall this close to the center line, suspect loss of sub-group integrity. Some major causes of loss of sub-group integrity are:

1. Over-control when used with serial sampling.
2. Two or more processes being mixed together and then sampled. Some examples are multiple suppliers, differing machines, and alternating shifts.
3. Some processes and controllers display a regular spike in the data or some other periodic pattern.

In all these cases, a plot of the individual data is a recommended starting point. Look for multiple systems by stratifying the data into potential causal groups (by shift, machine, supplier, etc).

In the previous example, the problem was over-control. The automatic process controller was "hunting." To illustrate this, a simple time series plot of the data

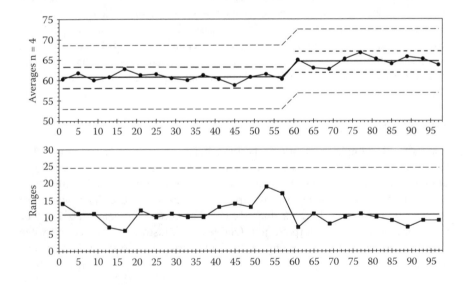

FIGURE 13.7 Over-control — average flow rate.

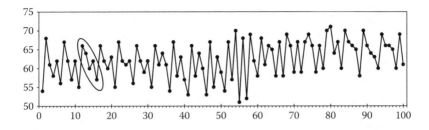

FIGURE 13.8 Over-control — flow rate for individual data.

is presented in Figure 13.8. Note the regular five-point pattern for the first half of the chart. Every group of five points in this region has one high and one low point, as well as three points in the mid-range region. (One five-point pattern is contained in an ellipse in Figure 13.8.) This is followed by a period where a saw-tooth pattern arises, followed by another five-point pattern that is so regular that it is linear. An immediate decision was made to turn the controller off and to run the flow rate in manual for one shift. The variation dropped to a third of the level noted in Figure 13.8.

The hugging of the center line in the averages chart was a consequence of nearly every sub-group having either a high or low point (or both) from the rhythmic cycle caused by the controller. This elevated all the range measurements, which in turn elevated $\bar{R}$. When $\bar{R}$ is inflated, so too is the spread of the control limits, resulting in the points giving the appearance of hugging the center line.

At the heart of the work of Shewhart, Deming, and the Six Sigma movement is the imperative of reducing variation. This is measured by the ranges chart. In no small way, success is measured in one's ability to understand and reduce variation, the evidence of which will be a lowered $\bar{R}$.

Any other occurrence that might cause regular loss of sub-group integrity has the potential to cause the points to hug the center line, particularly when serial sampling is in use. Some texts use this as a reason to recommend that averages charts not be used with serial sampling. However, if the chart is used with caution, an averages chart can still be useful in detecting over-control as well as slight but sustained shifts in the process mean. In this example, it achieved both. This led to substantial process improvement.

REFERENCES

1. W.A. Shewhart, *Statistical Method from the Viewpoint of Quality Control*, The Graduate School of Agriculture, Washington, D.C., 1939.
2. W.E. Deming, *Out of the Crisis*, MIT Press, Cambridge, MA, 1988.
3. D.J. Wheeler and D.S. Chambers, *Understanding Statistical Process Control*, SPC Press, Knoxville, TN, 1986.
4. E.L. Grant and R.S. Leavenworth, *Statistical Quality Control*, McGraw-Hill, New York, 1980.
5. D.J. Wheeler, *Advanced Topics in Statistical Process Control*, SPC Press, Knoxville, TN, 1995.

14 Origins and Theory

Some Theoretical Foundations

This chapter explains Shewhart's original experiments, illustrates how he determined limits, from where the tests for stability come, and the origins of the formulae used.

In order to illustrate his theory, Shewhart conducted extensive experiments using numbered chips that represented three different shapes of distributions. In each of three bowls, he placed wooden chips whose marked values formed a normal, triangular, and rectangular distribution, respectively. Figure 14.1 shows the authors' simplified distributions. Shewhart used distributions with 61 classes.[1] The simplified versions were developed to enable students to conduct the experiments in class more rapidly than would otherwise be the case.

After the chips were thoroughly mixed, individual samples ($n = 1$) were drawn from each bowl, with replacement. The chips are randomized between samples, so every chip has an equal opportunity of becoming a sample on each drawing. While sampling error is evident, it is no more than one would expect. No sample, regardless of size, can ever be a perfect representation of the population from which it was drawn. The $n = 1$ distributions formed are adequate representations of the contents of the bowls.

Then Shewhart grouped his data into sub-groups of four ($n = 4$) and made new distributions. During courses of instruction, the normal practice is to draw four chips instantaneously to save time. The probability of any given chip becoming a sample is altered by a tiny amount with this procedure, but this is not detectable in the data.

The distributions of the sample averages of $n = 4$ formed bell-shaped curves and approached normality in every case. The distributions of the sample averages are not exactly normal. They cannot be because no data will exactly fit a model as rigid as a normal curve, but their departure from normality is insignificant for the purposes of the experiment. Shewhart's experiments demonstrate that any shape of distribution, if stable, will yield a distribution that is bell-shaped provided the number of samples averaged is four or larger. The experiment is contrived. Provided the bowls are adequately stirred between samples and the sample taken is random, the data can be expected to demonstrate good stability.

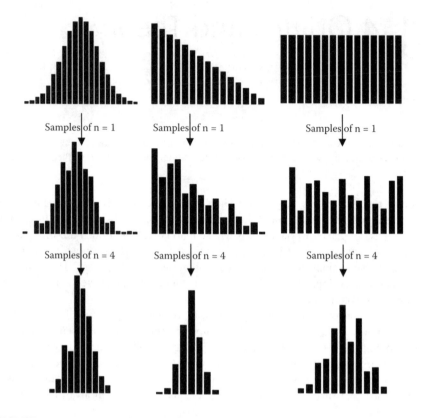

FIGURE 14.1 Results of a simplified Shewhart bowls experiment.

The experiment demonstrates that providing sample sizes of four or more are used:

1. If the process is stable, then the sample averages will always form a bell curve and will approach normality.

By extension,

2. If the sample averages are not bell-shaped, the process is not stable.
This is the third test for stability for an $\bar{x}$ and R control chart. The three tests are:

1. The plotted averages must fall at random about the center line.
2. They must remain within the control limits.
3. They should approximate a bell curve. That is, most points should fall reasonably close to the center line, with fewer and fewer points being found as the control limits are approached.

A glance at the charts in Figure 14.1 reveals two other characteristics that become important later in this chapter. First, the average of the sample averages ($\bar{\bar{x}}$) must always be the same as the average of the individual samples, within the limits of rounding. In addition, the range and standard deviation of the sample averages is less than it is for the individual samples. This follows intuition. There is only one chip marked with a 1 in the normal bowl. The probability that it would be sampled on four consecutive occasions is miniscule. If four chips are drawn at once, an average of 1 can never be obtained. The single chip marked with a 1 will be averaged with three other chips with higher values when the sample averages are calculated. The distribution of averages will always be narrower, meaning the range and standard deviation will be smaller.

DEVELOPING CONTROL LIMITS

Shewhart chose to place the control limits 3 $\sigma_{\bar{x}}$ on either side of the center line. This is extraordinarily convenient because it matches the general description of the normal curve. However, he chose these limits for economic rather than statistical reasons. Shewhart knew that it was possible to make both type one and type two errors, and he was attempting to minimize the net economic loss from both, as follows:

1. **Type one errors.** This describes the mistake made when a cause or explanation is sought for a data point that is in fact random. In legal terms, this is analogous to hanging an innocent man.
2. **Type two errors.** This is the mistake made when the data point in question is in fact "special"; it is a signal, but is erroneously assumed to be random. In this case, the murderer is set free, perhaps to kill again.

Society can ensure it never again convicts an innocent person by acquitting all accused persons. It can ensure it never lets a guilty person back on the streets by convicting every accused who appears in court. Instead, society chooses to minimize the impact of both types of mistakes, with the twin notions of the presumption of innocence and reasonable doubt.

Shewhart took a similar approach. He understood that quality is essentially an economic issue. Type one errors incur the cost of hunting for an imaginary problem, as well as the cost of developing erroneous cause-and-effect relationships. Type two errors allow causes of excessive variation to escape attention, ensuring that variability grows with time. The problem was to develop limits that produced minimum economic loss. Eventually, Shewhart settled for limits of 3 $\sigma_{\bar{x}}$ on either side of the center line This decision was based on empirical evidence and is guided by statistical theory. It has stood the test of time.

Shewhart mimicked the legal system by adopting the approach of assuming the data to be stable until a statistically significant signal indicated otherwise.

MAKING THE CONTROL CHART

As well as the $\bar{x}$ of each sub-group, Shewhart also calculated the σ of each sub-group. He then plotted the averages and sigmas separately, as seen in Figure 14.2.

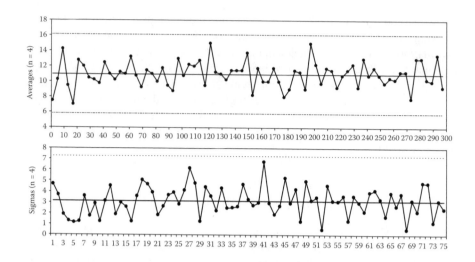

FIGURE 14.2 Shewhart bowl experiment $\bar{x}$ and σ control chart ($n = 4$).

As was noted in the previous chapter, it is possible to plot ranges instead of sigmas, but here it is necessary to return to the original form of the chart in order to illustrate the development of control limits, which is done in Appendix A of this chapter.

Having decided to place the control limits $3\sigma_{\bar{x}}$ on either side of the center line, Shewhart now decided how to calculate $\sigma_{\bar{x}}$. As discussed in Chapter 13, he knew that $\pm 3\sigma_{RMS}$ was not a workable option and needed an approach that gave a more reliable estimate of the limits of controlled variation; one that would not, for instance, be corrupted by over-control. The method he chose was to extrapolate from $\bar{\sigma}$ or from $\bar{R}$, depending on which approach was chosen. Readers who want to understand the origins of the formulae presented in Chapter 13 will find them in Appendix A of this chapter.

CONTROL LIMITS VARY WITH SUB-GROUP SIZE

One important characteristic of $\bar{x}$ and R control charts is that the spread of the control limits is a function of sub-group size. As the sample size increases, the spread of the control limits for the averages chart will shrink, as noted in Figure 14.1.

The absolute or known statistic of a population is indicated by the prime notation.[1] Therefore, the absolute population average is $\bar{x}'$, and sigma prime is represented by σ'. Of course, for control chart work, the absolute population parameters are never known. We sample to gain estimates of them.

During the Shewhart bowl experiments, it was noted that the distributions of the sample averages displayed less range, or a reduced spread, than did the individual values. If a sub-group size of eight were chosen, the sample averages would display an even lower range when distributed. There is a relationship between the degree

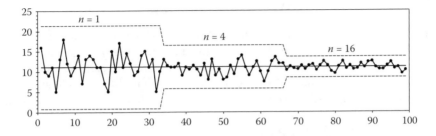

FIGURE 14.3 Control limits vary with sample size.

of variation displayed by sample averages and the sample size, and it is known as the standard error of the mean. It is expressed as:

$$\sigma_{\bar{x}} = \frac{\sigma'}{\sqrt{n}}$$

Therefore:

$$3\,\sigma_{\bar{x}} = \frac{3\sigma'}{\sqrt{n}}$$

As the sample size increases, $3\,\sigma_{\bar{x}}$ decreases. This results in reduction of the spread of the control limits as sample size grows. Figure 14.3 illustrates the change in control limits for sample sizes of 1, 4, and 16. Data from the authors' simplified Shewhart bowl experiment are used for this example.

SPECIFICATIONS AND CONTROL LIMITS

Because control limits are statistically calculated as the limits of controlled variation, they must never be confused with specifications. Shewhart's methods spring from a profoundly different view of variation than charts that show specifications. The traditional view of variation is a go, no-go perspective that separates good outcomes from bad. Shewhart's approach is to separate random from non-random, to separate "signals" from background "noise."

It is a fundamental error to draw specifications or tolerances on an $\bar{x}$ and R control chart. The control chart in Figure 14.3 illustrates why. Averages display less variation than do individual data. Suppose that for the data in Figure 14.3 the specifications were 5 and 15. Overlaying these specifications on the averages chart creates the illusion that data are meeting specifications, even when this is not the case. Specifications relate to individual data only.

WHY USE AVERAGES?

Given that Figure 14.3 demonstrates that it is possible to plot individual values rather than averages (charts for individuals are covered in Chapter 15), some are inclined to question the use of averages. Averages charts have several particular strengths, and three will be considered here: (1) the normalization of the sample averages, (2) sensitivity to slight drifts in the process mean,[2] and (3) the detection of over-control.

NORMALIZATION OF SAMPLE AVERAGES

As noted in Figure 14.1, sample averages always form a bell-shaped distribution if they are stable. Statistical theory demonstrates that it is not necessary for the sample averages to be even close to normal for Shewhart's formulae to work well.[3,4] The method and factors are sufficiently robust to accommodate significant departure from normality. Nonetheless, because stable sample averages will approach normality, they are symmetrical, making interpretation easier than would otherwise be the case (see Chapter 15). In addition, the normalized shape provides an important basis for the development of tests for stability, as will be seen later.

SENSITIVITY TO DRIFTS IN THE PROCESS MEAN

When there is a slight but sustained drift in the process mean, sample averages will detect this shift much more quickly than will a chart for individuals. There is a price to pay for this increased sensitivity, however. More data are needed to plot the same number of points per day or week as the chart for individuals. Nevertheless, averages are so much more sensitive in this respect as they offer a significant advantage. Figure 14.4 illustrates the increased sensitivity provided by averages. Charts for individuals and sample averages are compared. The data used in both cases comes from a Shewhart bowl experiment, but they have been modified to insert a $1 - \sigma$ shift in the process mean at sample number 145. Many people in the pharmaceutical industry would be inclined to declare the individuals chart to be stable. When a business is attempting to create Six Sigma levels of quality, it cannot often afford to be blind to changes of this magnitude.

DETECTION OF OVER-CONTROL

Chapter 13 dealt with this subject. Given the prevalence of over-control by automatic instruments in the pharmaceutical industry, it is an important subject. One real issue is that many controllers are programmed to use a variant of Rule 2 of the Nelson Funnel Experiment (Chapter 5). Under this circumstance, the data may exhibit a reasonably good state of control, despite the controller being responsible for an increase in variation. On some occasions, the best and fastest way to test a suspect controller is to turn it off and to control the process manually for a while. If variation drops, as is often the case, turn the controller back on. If variation rises again, this is better evidence of over-control than would have been obtained from any statistical technique. Steady-state trials that test controllers, as well as process set up and operation, regularly reduce variation in pharmaceutical processes by 40 to 60%.

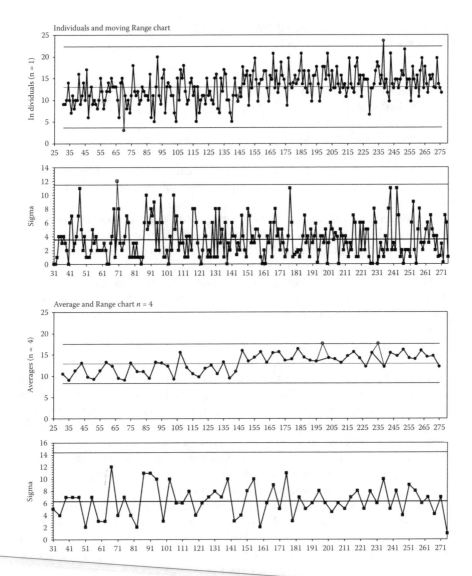

FIGURE 14.4 Averages chart superior at detecting slight but sustained shifts in the process mean.

Sometimes, this is all that is necessary to achieve Six Sigma levels of performance at several steps of the process. Figure 6.5, Figure 9.2, and Figure 13.5 are examples from earlier chapters that readers may choose to review.

INTERPRETING THE CHARTS

Earlier in this chapter, two important considerations were discussed. First, Shewhart's limits, while based on probability theory, are economic by nature and have an empirical element. Second, the essential tests for stability are that the plotted points for an $\bar{x}$ and R chart should:

1. fall at random about the center line
2. remain within the control limits
3. be bell shaped

Most points should fall reasonably close to the center line, with fewer and fewer points being found as the control limits are approached.

It is a mistake to extrapolate the rigid characteristics of the normal curve exactly to control charts. There are several reasons for this, some of which are discussed in the following.

The distribution of the averages is not exactly normal. The distribution is close to normal, and the closeness of fit increases as the sample size increases without bound, provided the data are stable. This places at least small limitations on how rigidly the normal law can be applied to control charts.

All processes waver, at least a little. The laws of thermodynamics guarantee as much. Bearings wear slowly, machinery degrades, mass and thermal balance never quite reach equilibrium in even the best-managed plants. Exact stability is not possible.

Some level of error is inevitable. Determining whether assignable causes are present requires inductive inference. Some level of type one and type two errors will always be present. Unfortunately, there is no way we can assign a precise probability to these events.[4,5] Every process has assignable causes of variation present. The evidence of this can be noted when a breakthrough in performance is achieved. Often, soon after the variation in a process has been substantially reduced, a new crop of assignable causes appears. They were always present. However, they remained invisible until the systematic variation was reduced. These "hidden" assignable causes are included in the calculations for the control limits, altering the likelihood of type one and type two errors, but to an unknown degree.

Therefore, the tests for stability, while built on a foundation of sound theory, are partially empirical and inevitably have an inbuilt error rate. Scientifically inclined people, such as are found in the pharmaceutical industry, want to see a demonstration of from where the tests for stability come. The following paragraphs will use an approach that cannot be exact because inductive inference precludes any exactitude, but which demonstrates that the tests are based on statistical theory.

To indicate how the tests are arrived at, consider the level of error that is known to be inevitable. It is about 3 in 1000. Even if the averages plotted were perfectly normal and stable, because the limits are placed 3 $\sigma_{\bar{x}}$ on either side of the center line, approximately this proportion of the data will fall beyond the limits in accordance with normal law. Three in 1000 is approximately the same as 1:333. An empirical adjustment must be made to allow for the issues discussed earlier in this section, and for illustrative purposes, this inbuilt error rate will be empirically adjusted to 1:250.

If that level of error is used, then any event that has a probability of happening that is less than 1:250 will be deemed to be statistically significant, and an assignable cause should be sought.

For instance, the probability of any point falling on one side of the average of a stable chart is $1/2$. The probability that eight consecutive points will fall on one side of the center line in a stable process is:

$$\frac{1}{2} \times \frac{1}{2} \times \frac{1}{2} \times \frac{1}{2} \times \frac{1}{2} \times \frac{1}{2} \times \frac{1}{2} \times \frac{1}{2} = \frac{1}{256}$$

This is a lower probability than 1:250, and the process is deemed to be unstable. Something changed. A review of the literature reveals less than perfect agreement on tests for stability for control charts. In this simple case alone, opinions vary. Deming suggested eight consecutive points indicated instability,[6] Ishikawa chose seven,[7] Juran opted for eight,[1] and Kume used seven.[8]

The fact that such luminaries do not agree perfectly is a function of the empirical element that the tests for stability contain. Nevertheless, the difference between established authors is never enormous, and a common theme is clear.

From normal law, the probability that a point will fall randomly between 2σ and 3σ on a given side of the center line is about 2%. Therefore, the probability that two consecutive points will randomly fall between 2σ and 3σ on one side of the center line is $1/50 \times 1/50 = 1:2500$. This is 10 times more powerful a test of one point falling beyond control limits. In fact, this test is so powerful the points do not need to be consecutive. Two out of three points falling between 2σ and 3σ and on the same side of the center line is evidence enough that something has changed.

The authors' preferred tests and guidelines follow.

TESTS FOR STABILITY

The following tests for stability, if noted, indicate non-randomness; an assignable or special cause has impacted the data; something has changed:

1. One or more points fall beyond the control limits.
2. Two consecutive points fall between 2σ and 3σ from, and on the same side of, the center line.
3. Two out of three points between 2σ and 3σ from the center line, where all three fall on the same side of the center line.
4. Seven or eight (choose one to use company-wide) consecutive points fall on one side of the center line.

GUIDELINES FOR INVESTIGATION

Further guidelines, rather than tests, are:

1. Any regular and repeatable pattern is an indication of non-randomness (for examples, see Figure 11.2 and Figure 13.10). A common example is the saw-tooth pattern of over-control. Strictly speaking, 14 points should alternate up and down to fail a test, but there is room for the use of

judgment. Periodicity is another example of a repeatable pattern. Figure 11.5 provides an example.

2. If the bulk of the points fall within 1 σ of the center line, non-randomness is suggested and this condition should be investigated. Again, some room for judgment exists, as a single assignable cause can break this pattern, but that would not be grounds to ignore it.

3. If the bulk of the points fall near the control limits, and very few fall near the center line, the data is not bell-shaped and the process should be investigated. Over-control is one possibility. The presence of multiple processes is another.

The literature contains many more potential tests, but these are unnecessary and sometimes counterproductive. As the number of tests and guidelines increase, so too does the probability of a false alarm. The authors' experience is that tests 1 to 4, together with the guidelines, are sufficient until the process becomes extremely robust, when other tests may be considered.[4] In fact, anything more than the recommended number of tests and guidelines is something of a moot point for most businesses. Many plants in the pharmaceutical industry would be overjoyed if only most of the points would remain within properly computed control limits.

THE FINAL WORD

The final word belongs to Shewhart, who wrote:

> The long range contribution of statistics depends not so much upon getting a lot of highly trained statisticians into industry as it does in creating a statistically minded generation of physicists, chemists, engineers and others who will in any way have a hand in developing and directing the production processes of tomorrow.[9]

Engineers, chemists, and biologists all bring their technical skills to any study of the process. Shewhart's objective was not to supplant these skills. Rather, it was to supplement them. He believed that statistically minded technical people were far more empowered to improve the process than those who lacked basic statistical thinking. In this sense, it is wise to study the process in technical as well as statistical terms.

For instance, if a chemist sees the data point for the last batch is only just inside limits, but on a technical basis she suspects that there may have been a process upset, she would be foolish to ignore the technical concerns merely because the point for that batch was barely inside the limits.

If a saw-tooth pattern appears to be emerging, should not the responsible people investigate, rather than wait for 14 points to alternate up and down without a break in this pattern? Sometimes statistical methods provide clues rather than clear signals.

If a technical clue indicating potential trouble were found, would it not be wise to study the processes implicated by this clue on a statistical as well as technical basis?

The people who achieve most tend to be those who successfully blend their technical knowledge with sound statistical thinking. For example, the engineer who understands why rational sub-grouping is important (see Chapter 15) is likely to learn more about the process than one who does not.

The engineers, chemists, and biologists who, guided by statistical theory, are better able to ask the right questions, conduct statistical investigations, and follow up on statistical signals and clues will be rewarded with reduced variation, improved productivity, and a business where people take pride and joy in their work.

REFERENCES

1. J.M. Juran, F.M. Gryna, and R.S. Bingham, *Quality Control Handbook*, McGraw-Hill, New York, 1974.
2. E.L. Grant and R.S. Leavenworth, *Statistical Quality Control*, McGraw-Hill, New York, 1980.
3. W.A. Shewhart, *Economic Control of Quality of Manufactured Product*, Van Nostrand, New York, 1931.
4. D.J. Wheeler, *Advanced Topics in Statistical Process Control*, SPC Press, Knoxville, TN, 1995.
5. W.E. Deming, On probability as a basis for action, *The American Statistician,* 29(4), 146–152, 1975.
6. W.E. Deming, *Out of the Crisis*, MIT Press, Cambridge, MA, 1988.
7. K. Ishikawa, *Guide to Quality Control*, Asian Productivity Organization, Tokyo, 1974.
8. H. Kume, *Statistical Methods for Quality Improvement*, 3A Corporation, Tokyo, 1985.
9. W.A. Shewhart, *Statistical Method from the Viewpoint of Quality Control*, The Graduate School of Agriculture, Washington, D.C., 1939.

Appendix A

Origins of the Formulae

Some Additional Detail

THE AVERAGES CHART

Control limits are placed three standard deviations of the averages on either side of the center line.

For the averages chart:

$$\text{U/LCL} = \overline{\overline{x}} \pm 3\,\sigma_{\overline{x}}$$

$$= \overline{\overline{x}} \pm \frac{3\sigma'}{\sqrt{n}}$$

It is known that $\overline{\sigma} \neq \sigma'$. Large samples tend to expose more of the population variation than do small samples. In the bowls experiment, small samples would tend to have lower values for σ than would large samples. As the sample size increased, the estimate for σ would improve and approach the population parameter. The relationship between $\overline{\sigma}$ and σ' is noted in Figure 14.5.

Shewhart then demonstrated that if σ' was stable, and if a long enough series of sub-groups was used, the relationship between $\overline{\sigma}$ and σ' was:

For any given n,

$$\frac{\overline{\sigma}}{\sigma'} = C_2$$

By extension:

$$\sigma' = \frac{\overline{\sigma}}{C_2}$$

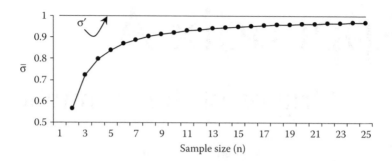

FIGURE 14.5 Relationship between $\bar{\sigma}$ and σ' (where $\sigma' = 1$).

For the averages chart:

$$U/LCL = \bar{\bar{x}} \pm \frac{3\sigma'}{\sqrt{n}}$$

$$= \bar{\bar{x}} \pm \frac{3\bar{\sigma}}{C_2\sqrt{n}}$$

Because 3, C_2, and n are constant for any given sample size:

For any given n, let the factor: $A_1 = \dfrac{3}{C_2\sqrt{n}}$

For the averages chart: $U/LCL = \bar{\bar{x}} \pm A_1\bar{\sigma}$

Simplifying the arithmetic. It is possible to substitute range for sigma as a measure of variation in a control chart. For sample sizes up to about 8, range is a superior estimator, after which time sigma becomes a better estimator.[1]

Given that there exists a relationship between $\bar{\sigma}$ and σ', it is no surprise to discover that a similar relationship exists between $\bar{R}$ and σ'. That relationship is expressed by the d_2 factor and is:

$$\frac{\bar{R}}{\sigma'} = d_2$$

By extension:

$$\sigma' = \frac{\bar{R}}{d_2}$$

Given that:

$$U/LCL = \overline{\overline{x}} \pm \frac{3\sigma'}{\sqrt{n}}$$

$$= \overline{\overline{x}} \pm \frac{\overline{R}}{d_2\sqrt{n}}$$

Because 3, d_2, and n are constant for any given sample size:

For any given n, let the factor $A_2 = \dfrac{3}{d_2\sqrt{n}}$

For the averages chart: $U/LCL = \overline{\overline{x}} \pm A_2 \overline{R}$

THE SIGMA CHART

Shewhart expressed the relationship between σ_σ and $\overline{\sigma}$ using the C_2 factor. From distribution theory:

$$\sigma_\sigma = \frac{\sigma'}{\sqrt{2n}} = \frac{\overline{\sigma}}{C_2\sqrt{2n}}$$

For the sigma chart:

$$U/LCL = \overline{\sigma} \pm \frac{3\overline{\sigma}}{C_2\sqrt{2n}}$$

However, theory demonstrates that sigma charts behave in a manner similar to attributes charts. There is a relationship between $\overline{\sigma}$ and the control limits, which allows a direct extrapolation from $\overline{\sigma}$ to the control limits. The B_3 and B_4 factors express this relationship.[1] Using them, the algebra for control limits can be simplified as:

$$UCL_\sigma = B_4\,\overline{\sigma}$$

$$LCL_\sigma = B_3\,\overline{\sigma}$$

THE RANGE CHART

Statistical theory defines the expected distribution of R when sampling from a normal population. In addition, theory gives the ratio of $\bar{R}$ to σ'.[1] The ranges chart behaves in a manner similar to sigma charts. Because of this, it is possible to extrapolate directly from $\bar{R}$ to the control limits. The formulae are:

$$UCL_R = D_4\,\bar{R}$$

$$LCL_R = D_3\,\bar{R}$$

FACTORS

All factors discussed in this appendix can be found in Appendix 1, Appendix 2, or Appendix 3 at the end of this book.

REFERENCES

1. E.L. Grant and R.S. Leavenworth, *Statistical Quality Control*, McGraw-Hill, New York, 1980.
2. J.M. Juran, F.M. Gryna, and R.S. Bingham, *Quality Control Handbook*, McGraw-Hill, New York, 1974.

15 Charts for Individuals

Individual Point and Moving Range Charts

Sometimes it is neither desirable nor practical to form sub-groups and to construct an average and range control chart. On such occasions, an individual point and moving range chart may be used.

The individual point and moving range control chart allows data collected by serial sampling to be plotted individually, as is normally the case for serial sampling. With the exception of the methodology for calculating ranges, its construction is very similar to average and range control charts. Figure 15.1 shows 56 modified data from the Red Beads Experiment. A reduction in variation was simulated by halving the variation around the center line from the 25th point onward. At point 40, the process average was shifted down by 3σ. Special causes of similar magnitude were applied at points 12 and 34. Red beads data are used here because they are known to be stable, and they illustrate how the chart demonstrates changes.

The charts in Figure 15.1 show three populations. There is a shift in mean for both charts, but at different times. This behavior is a function of variables data, and is one reason why there are two charts for variables to track central tendency and variation separately. Two important points are worth noting before moving on to how the charts are constructed:

1. **Special causes.** Note how the second special cause is more apparent, despite being of similar magnitude to the first. This is a function of the reduced background variation. Reduced background noise will always make signals easier to find. Furthermore, each special cause in the individual chart corresponds with two out of control points in the moving ranges chart. These points should be removed from the calculations from the moving $\bar{R}$ to avoid inflating the control limits.

2. **Shift in process mean.** When there is a shift in the process mean, there is likelihood that there will be a corresponding special cause in the ranges chart. This point should be removed from the calculations for the moving $\bar{R}$.

CONSTRUCTING THE CHARTS

The first 14 samples for the chart in Figure 15.1 are in Table 15.1. The moving range is arrived at by calculating the difference between successive values. Note that under the first datum the moving range has no value.

SampleNo.	1	2	3	4	5	6	7	8	9	10	11	12	13	14
Data	27	26	21	29	25	27	26	27	23	28	25	38	22	25

SampleNo.	15	16	17	18	19	20	21	22	23	24	25	26	27	28
Data	25	31	27	26	29	25	21	27	25	30	24	27	26	26

SampleNo.	29	30	31	32	33	34	35	36	37	38	39	40	41	42
Data	29	26	26	25	28	12	27	26	23	26	27	21	22	22

SampleNo.	43	44	45	46	47	48	49	50	51	52	53	54	55	56
Data	21	23	20	22	22	22	20	23	21	24	21	20	21	22

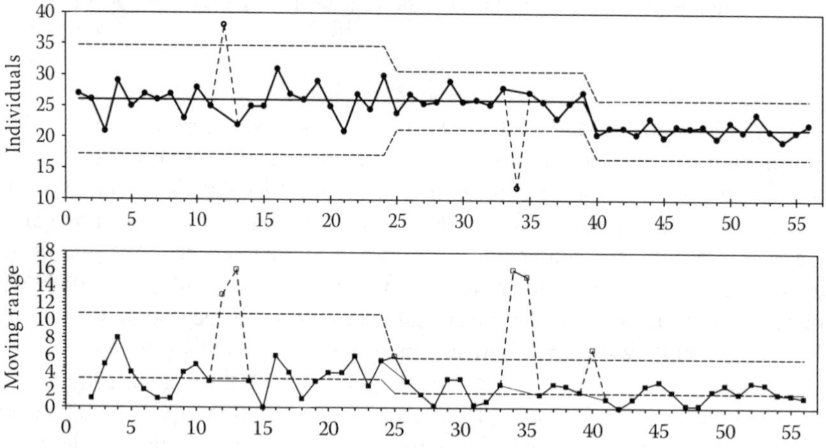

FIGURE 15.1 Data and individuals chart — modified red beads data.

The formulae are similar to average and range charts, as follows:

Individuals chart: $U/LCL = \bar{x} \pm A_2 M \bar{R}$ (For $n = 1$, $A_2 = 2.66$)

Moving R chart: $UCL = D_4 M \bar{R}$ (For $n = 1$, $D_4 = 3.27$)

$$UCL = 0$$

TABLE 15.1
Data for Individual Point and Moving Range Control Chart

Sample No.	1	2	3	4	5	6	7	8	9	10	11	12	13	14
Data	27	26	21	29	25	27	26	27	23	28	25	38	22	25
Moving Range	X	1	5	8	4	2	1	1	4	5	3	13	16	3

For the first population:

Individuals Chart:	Moving Range Chart:
$U/LCL = \bar{x} \pm A_2 M\,\bar{R}$	$UCL = D_4 M\,\bar{R}$
$= 26 \pm 2.66 \times 3.32$	$= 3.27 \times 3.32$
$= 26 \pm 8.8$	$= 10.9$
$UCL = 34.8$	
$LCL = 17.2$	

For the second population:

Individuals Chart:	Moving Range Chart:
$U/LC = \bar{x} \pm A_2 M\,\bar{R}$	$UCL = D_4 M\,\bar{R}$
$= 26 \pm 2.66 \times 1.76$	$= 3.27 \times 1.76$
$= 26 \pm 4.6$	$= 5.7$
$UCL = 30.6$	
$LCL = 21.4$	

For the third population:

Individuals Chart:	Moving Range Chart:
$U/LCL = \bar{x} \pm A_2 M\,\bar{R}$	$UCL = D_4 M\,\bar{R}$
$= 21.4 \pm 2.66 \times 1.76$	$= 3.27 \times 1.76$
$= 21.4 \pm 4.6$	$= 5.7$
$UCL = 26.0$	
$LCL = 16.8$	

The charts are then drawn in the same manner as $\bar{x}$ and R control charts.

INTERPRETING INDIVIDUAL POINT AND MOVING RANGE CHARTS

Individual point data can be any shape or functional form; not all are bell-shaped. This limits the use of the normal law to interpret the charts. Thankfully, the empirical rule comes to the rescue.[1] The empirical rule can be applied to distributions of any shape or functional form, regardless of whether they are symmetrical or not. This rule states:

Given a homogeneous set of data:

1. Approximately 60 to 75% of the data will fall $\pm 1\sigma$ of the mean.
2. Approximately 90 to 98% of the data will fall $\pm 2\sigma$ of the mean.
3. Approximately 99 to 100% of the data will fall $\pm 3\sigma$ of the mean.

Batch No.	1	2	3	4	5	6	7	8	9	10	11	12	13	14	15	16
Data	12	48	9	9	28	17	14	19	19	5	13	12	5	50	47	19

Batch No.	17	18	19	20	21	22	23	24	25	26	27	28	29	30	31	32
Data	87	5	8	49	36	5	54	24	27	41	73	10	23	5	17	42

Batch No.	33	34	35	36	37	38	39	40	41	42	43	44	45	46	47	48
Data	59	5	20	75	5	25	7	5	5	93	72	5	31	5	19	5

Batch No .	49	50	51	52	53	54	55	56	57	58	59	60	61	62	63	64
Data	52	19	18	5	33	26	5	5	84	5	21	61	11	34	5	65

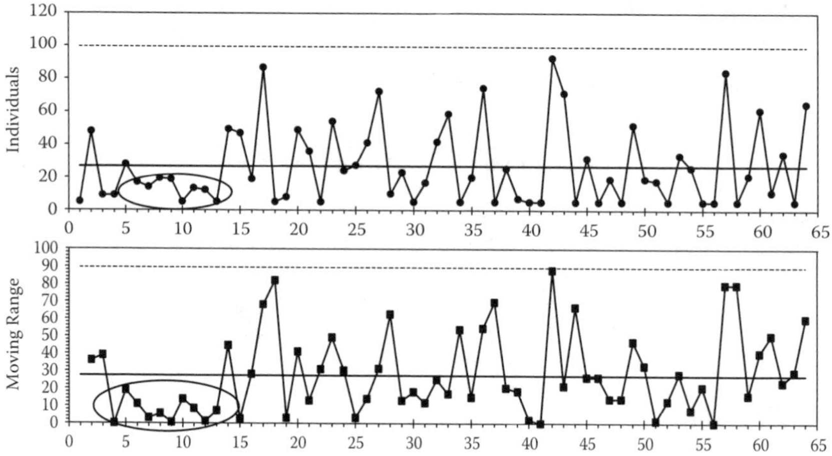

FIGURE 15.2 Impurities — individual point and moving range control chart.

Of course, with data that have a significant departure from normality, the distribution will not be symmetrical. When this occurs, the tests for stability used for average and range charts cannot be applied with similar confidence to the individuals chart. An example of this nonsymmetry can be found in Figure 15.2 and Figure 15.3. These charts are for the level of impurity in a therapeutic compound, measured in parts per million (ppm). In this case, the detectable limit is 5 ppm. Note the lack of symmetry in the individuals chart. The LCL is not visible because it was calculated to be below zero.

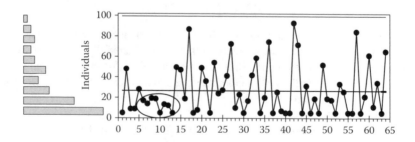

FIGURE 15.3 Distribution and individuals chart — impurities.

To allow a direct comparison between the plot of individuals and the frequency distribution for the same data, both are shown in Figure 15.3. The distribution is anything but normal, having a functional form that appears exponential.

If the data are symmetrical around the average, the tests for interpretation used for average charts may be used for the individuals chart. If the data are non-symmetrical, is this due to the natural shape of the distribution, or is it a function of special causes, all of which fall on one side of the center line? The moving ranges chart can help here. If points fall beyond the control limits, remove them from the calculations and study the individuals chart again.

Usually, special causes are easier to discern in a moving range chart than they are in an average and range chart. This is because of a characteristic pattern formed. Whenever there is a special cause in the individual chart, two consecutive high points will be seen in the moving ranges chart. Two consecutive high points (both points more than 2σ from center line) in the ranges chart is almost always an indication of a special cause in the individuals chart. The exception is when exponentially (or similarly) distributed data are involved, as noted in Figure 15.3.

A single out of control point in the moving ranges chart is an indication that the process mean of the individuals chart has shifted (see Figure 15.1). However, if the process mean shifts by a only a small amount, or drifts slowly to a new level, as opposed to a sudden step, no out of control point will be found in the ranges chart.

The literature suggests that the run tests (number of consecutive points on one side of the center line) should not be used with moving range charts. To test this notion empirically, several hundred samples were taken from the Shewhart bowl experiments using the authors' equipment. Two moving range charts for 200 data each are shown in Figure 15.4.

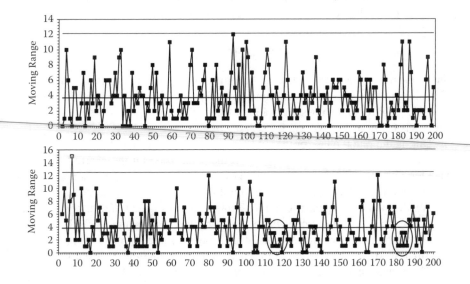

FIGURE 15.4 Two moving ranges charts — Shewhart bowl experiment.

The top chart shows no irregularities. The bottom chart shows two occasions where seven or more points fell below center line (circled). The evidence supports the literature. Splitting the control limits for the moving range chart to indicate a change in the center line should be done with caution. Generally, it is best to show a shift in the moving range ($M/\bar{R}$) only if one is sure that there has been a change made to the process; that is, one should be able to use a combination of technical and statistical knowledge. An example would be to split limits at the time of a plant trial. Another example would be after procedures had been changed or tightened up with a view to reducing variation, in which case the purpose of the chart would be to determine whether the change had been effective.

SUMMARY

The tests for stability for an individual point and moving range control chart are:

1. **If the individual data appear symmetrical,** use the same tests for the individuals chart as recommended for the averages chart.
2. **If the individual data appear nonsymmetrical,** a judgment call is necessary. One can choose to use the tests as per the averages chart, knowing that more type one errors are likely, or one can choose not to act without technical information that supports the idea that a signal has been found. In the authors' experience, the former approach is adopted in most cases, followed by a technical investigation to verify the signal, and to determine its nature.
3. **Moving range charts.** Unless technical evidence supports the notion that the moving $\bar{R}$ has shifted, do not recalculate $\bar{R}$.
4. **Guidelines.** Use the same guidelines as per the average and range charts. In particular, look for pattern and shape in the data.

STRATIFICATION

One of the important aspects of control charting already mentioned is the importance of stratifying data into shifts, machines suppliers, or any other technically rational grouping. Figure 15.5 shows an example where the initial dose from an automatic injector is being measured.

During the initial phase of the study, one team member thought he saw a pattern in that nearly all the high points were in the same zone on the chart. This could have been natural variation, or it could have been the consequence of a non-normal distribution, but a decision was made to follow up on the idea.

Another team member, a QA representative, investigated and discovered that there were two facilities from where the cartridges were sourced. Both facilities were using the same specifications and were using the same suppliers of components; nevertheless, the data were sorted to make two charts, one for each supplier.

The resultant charts are shown in Figure 15.5, below the initial chart. The difference in performance of the two suppliers is evident. Whenever there are two shifts, suppliers, machines, or instruments, one can start with the assumption that they are different.

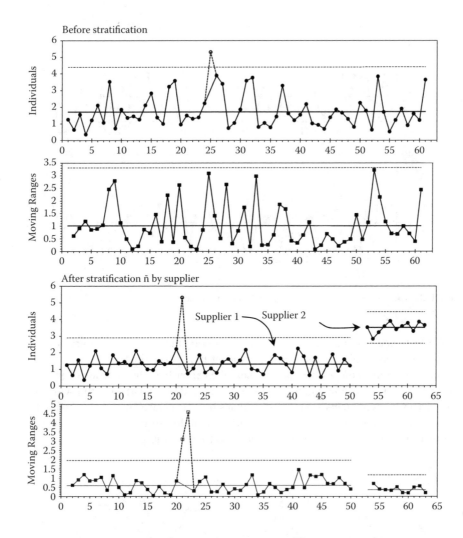

FIGURE 15.5 Automatic injectors — key performance function.

It is necessary to know in what way are they different, and by how much. In this case, the variation could be enough to make the difference between products passing or failing final release tests. The next phase of any study such as this is to determine whether to work at the suppliers' facilities to make the two products more alike, or to make the downstream process robust to the differences between the facilities.

PATTERN AND SHAPE

The importance of looking for pattern and shape in the data has been emphasized several times. The chart in Figure 15.6 is an example where the correct interpretation of the chart is a function of seeing pattern and shape. These data are a measure of

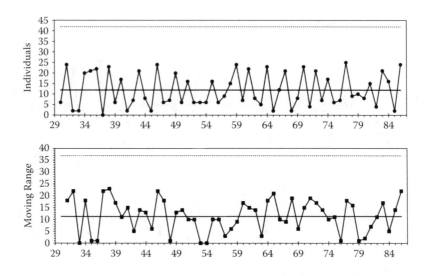

FIGURE 15.6 Effectiveness of a chemical bonding process.

the effectiveness of a protein-folding step involving the formation of intramolecular disulfide bonds. This process has a significant impact on yield.

Note how almost none of the data fall near the center line, the opposite of what one would expect. Also, the control limits in the individuals chart look too wide, as if the ranges measurements are inflated, perhaps by over-control, or perhaps by multiple processes being blended on a single chart. The ability to find pattern and shape is critical. In this case, it provided the clues that led to a large improvement in yield

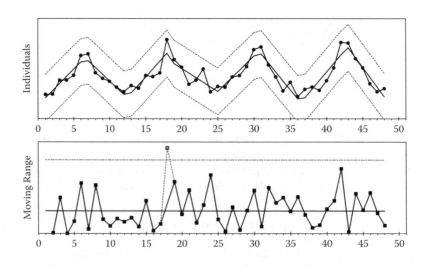

FIGURE 15.7 Periodicity in data.

when all operators across all shifts were calibrated so they all performed each operation the same way. In addition, real time process control using Shewhart's methods was introduced (see Chapter 16). The pattern in the chart in Figure 15.6 was caused by between-operator and between-shift variability.

The danger is that people will transfer their old go, no-go thinking from charts that showed specifications to Shewhart control charts. Such a sub-optimal approach can be like chasing nickels and dimes from a dropped wallet, while the $20 bills drift away in the breeze.

PERIODICITY

Some processes show a regular rise and fall pattern. Temperature often displays this characteristic. The example in Figure 15.7 shows monthly sales for an upper respiratory tract treatment (seen earlier in Chapter 11). The annual cycle in the data is evident, but for any given time of the year, the data are predictable and the process is stable. Because the process is continuously trending, the moving ranges will tend to be inflated by this movement of the system. For charts with a trending center line, the control limits can be expected to be at least slightly further apart than would otherwise be the case.

REFERENCE

1. D.J. Wheeler, *Advanced Topics in Statistical Process Control*, SPC Press, Knoxville, TN, 1995.

16 Practical Considerations

Some Additional Concepts and Applications

Here we introduce the reader to some practical applications. Process control, improving understanding of causal relationships, and some additional statistical concepts as well as a "how to" guide are discussed.

WHAT DO THE STATISTICS MEAN?

What information do statistics such as average, standard deviation, skewness, or kurtosis mean? Sometimes they mean very little. This will be the case whenever instability is found.[1]

By definition, if the process is unstable, it is unpredictable and therefore has no known capability. The calculations can be done, and resultants will be forthcoming, but they mean very little. The standard deviation of an unstable process will alter depending on how many special causes or shifts in the process mean are included in the data set. It is a matter of luck.

Stabilize first *is* the first principle in development as well as in production processes and laboratories. If for no other reason than unstable data give erroneous, non-repeatable resultants for process average, standard deviation, and process capability, stabilize first *must* be the first principle. If the data lack meaning, there seems to be little point in conducting any calculations, let alone any interpretation.

Suppose during the development phase of a new drug that data are gathered on critical process parameters. It is normal practice to calculate the average and standard deviation of the data during development, and then to compute limits three standard deviations on either side of the center line. These become the basis for the specifications or standards against which future production performance is measured.

Several months later, after full production has been achieved, the production data are examined against the standards and specifications. How likely is it that the new numbers generated will be close to the development data? Whatever the process capability study indicates, does it have any meaning? If the data are unstable, in all likelihood significant discrepancies will exist. This leads to wondering which set of data are more reliable, blaming the additional variation on the difference between a small-scale development process and a full-scale production process, or conducting still more tests in an attempt to "firm up" the analysis. All of these approaches are short roads to Hades.

Production people everywhere pray for a few upsets in the development phase. The greater the number of special causes that affect the development process, the wider will be the standards and specifications when they are transferred to production. Production specifications and compliance should depend more on science than on luck.

Development data are difficult to handle because they contain short- and perhaps medium-term variation, but no long-term variation. Therefore, a full-scale production process is nearly always going to show more variation than the development data suggest.

The compliance implications, in production as well as in the laboratory, are obvious. If the initial standards set up under short-run development conditions are not adjusted to take into account the inevitable additional medium- to long-term variation, the production process will almost certainly fail to routinely meet these standards. Deviations will gradually increase and investigations will soak up more and more time, creating even more pressure.

RATIONAL SUB-GROUPS

A critical part of control chart theory and practice for the average and range control chart is the creation of rational sub-groups.[1-3] The most common definition for this term found in the literature is that all samples in any given sub-group should come from "the same essential technical conditions." Put another way, there should be no changes to the process *within* the sub-group. This helps to ensure sub-group integrity and to produce a reliable estimate of $\bar{R}$. However, Shewhart noted that the phrase "the same essential technical conditions," can become something of a secret code passed between people where each assumes the other knows what he is talking about, but where neither does.[1] As has been discussed earlier, it is the within-sub-group variation, illustrated by the ranges chart, that determines the spread of all control limits. The guidelines in the following paragraphs are suggested.

Guideline One. Ensure sampling and test procedures are stable with minimum variation. In the pharmaceutical industry, a good place to start statistical investigations and process improvement is sampling and the analytical processes. If sampling and analytical error are high, detecting signals from the production data will be more difficult than need be the case. If the sampling and analytical processes are unstable, the signals can send the production people looking for problems in the plant that are in fact part of the sampling and analytical processes.

Guideline Two. Minimize the opportunity for variation *within* each sub-group:

1. **Group like with like**. The approach to sub-grouping should attempt to minimize the variation *within* each sub-group. This keeps the ranges measurements lower and makes the chart more sensitive to signals in the averages chart.
2. **Stratify the data into separate charts**. If you have more than one instrument, machine, fermenter, cavity, filling head, or shift, log the data so it can easily be sorted and then plotted by these characteristics. There is nothing inherently wrong with combining the data from multiple machines or shifts to get an overall readout on the process, *provided that each of these things is always plotted and analyzed separately so the unique contribution to variation of each source is understood.* Figure 11.2 and Figure 15.6 provide examples. Reference 2 has several other excellent examples.

Guideline Three. Potential changes to the process should occur between sub-groups for averages charts. If you suspect a change to the process, or any other signal, ensure that all the data in one sub-group do not include this change, and all the data in the next sub-group do contain the change. This is particularly important when serial sampling is being used in an average and range control chart.

THE BLESSING OF CHAOS

The worst thing one can see when preparing the first control charts for a process is reasonable stability. Fortunately, this is rare in the pharmaceutical industry. Chaos is a blessing; it is what Six Sigma practitioners should hope will jump off the page at them from their initial control charts. If the process exhibits chaos, then if one does nothing more than stabilize the process, significant improvements will follow; no changes to technology are required, GMP procedures may remain as they are, eliminating any immediate need to process change controls. Sometimes, doing nothing more than stabilizing the process creates Six Sigma levels of quality. Always, it reduces variation significantly if the process is in a state of chaos at the outset.

STABILIZING A PROCESS

THE BRUTE FORCE APPROACH

A review of the literature reveals that the most common advice given to those needing to stabilize a process is to find and eliminate the special causes of variation. If the initial charts reveal a baseline of systematic variation with occasional assignable causes, this approach will do the job. However, if the process is in chaos, or is approaching a state of chaos, such a methodology often fails. Any progress made is slow and exhausting. In these situations, the brute force approach is a fast and effective way to bring the process into control, at or approaching Six Sigma levels of variation.

An example of this approach was provided in the first case study in Chapter 9. Readers are invited to review this case study before proceeding. The brute force approach aims at driving every variable to as low a level as possible as part of a formally organized, planned, and executed operational directive. In the pharmaceutical industry, there is a tendency to target the so-called "critical parameters" only. If Six Sigma levels of quality are aimed for, this is a poor approach. In any event, it is a half-hearted, dismal approach. The interactive nature of the variables in chemical and biological processes means that a parameter thought to be non-critical can interact with other variables to increase variability in a critical parameter. The enemy is variation, all variation, not just some of it. If a business is going to the trouble of conducting a steady-state trial or similar test, it may as well make all the improvements possible and maximize the return on time invested.

In the example of a steady-state trial given in Chapter 9, there was an unfortunate epilogue. Although significant improvements were made and locked in by reprogramming the controller, the improvements made by creating uniform operating procedures across all shifts were diminished soon after the trial. The problem was that the activity was called a trial. Once the end date for the trial had passed, some people relaxed their vigilance. No long-term control measures were put in place, and some variation crept back into the process. The "C" in DMAIC (Define, Measure, Analyze, Improve, and Control) stands for control, and this element was not effectively pursued.

PROCEDURE — THE BRUTE FORCE APPROACH

An outline Operational Directive from a fictional pharmaceutical company appears in Appendix A to this chapter, as a further guide to operationalizing the brute force approach. The major components appear in the following paragraphs.

Define. The problem is defined as instability or high variation. The aim is to both stabilize the process and to drive variation to a minimum, to determine the underlying process capability, and to expose causal relationships. Note that the aim is not to improve yield, nor is it to reduce costs. These objectives come after minimum variation has been achieved.

Measure. Measure and plot on control charts all key inputs, outputs, and process parameters to form a baseline for the project. Look for signs of high variation and instability so these areas can be targeted.

Analyze. For a project such as this, the analysis phase is less about analyzing data than it is about analyzing the way the work is done. Is there variability between shifts? Is the level of skill among operators adequate? How will managers and supervisors ensure uniform operating procedures across all shifts? Does the creation of process owners for each technology type seem appropriate? Are automatic process controllers involved? How will they be checked for over-control? Do chemists routinely alter formulations based on recent results? Are the laboratory controls stable? How can the between-analyst and between-instrument variation be minimized? Are sampling procedures uniform? How do we know they are uniform? Is maintenance of equipment up to date? How many suppliers of raw materials and components exist? How often do batch changes in raw materials occur? Is it possible to use only one working cell bank to inoculate all fermenters, at least initially? An analysis to discover key variables (see section "What to Measure and Plot") will add to this list of questions. Every potential source of variability must be identified.

Improve. In this type of example, perhaps **implement** would be a better word. After careful planning and briefing of all staff, a formal operational directive is issued. The process is driven into a state of control with minimal variation to achieve excellent precision or repeatability.

Control. Once the process is stable and operating with minimum variation, wherever possible control measures should be built into the way the work is done. A cultural aversion to variability can play a vital role in ensuring variation is not allowed to drift back into the process. The creation of such a culture is a leadership

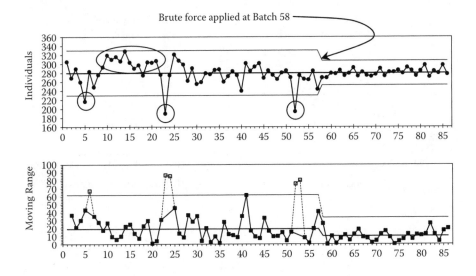

FIGURE 16.1 Specific activity before and after brute force was applied.

issue. Charts for process control, discussed in the section titled "Process Control," are an important component of the control element.

CASE STUDY

In this example, a preliminary study indicated high variation throughout the entire process. A decision was made to drive the earliest events into control with minimum variability. The first stage of the process was fermentation, where bacteria were grown to provide the cellular material that formed the basis for the therapeutic compound produced.

The initial statistical analyses indicated that the process was afflicted by assignable causes, but it did not appear to be in a state of chaos. A typical chart, for Specific Activity, appears in Figure 16.1.

Several weeks were taken to prepare to execute the plant manager's operational directive. At Batch 58, the directive was enacted. The results were excellent. Variability more than halved. The sigma of the data up to Batch 57 was 27.24. The sigma from Batch 58 onward was 10.21. As will be seen in the next section, this uniform operation was only the beginning of the benefits gained. The low levels of variability achieved clarified causal relationships and led to further process improvement studies.

CAUSAL RELATIONSHIPS

The difficulty experienced by technical people in establishing casual relationships in the pharmaceutical industry has been discussed several times. When the process is unstable, identifying causal relationships can be difficult; quantifying them can be a nightmare. However, reasonable stability and low variation can be of real

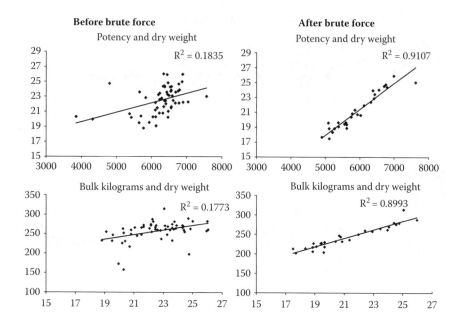

FIGURE 16.2 Correlations before and after brute force directive.

assistance in both respects. To illustrate, Figure 16.2 shows some correlations before and after the brute force approach was applied to the fermentation process discussed in the previous part of this chapter.

Initially, potency and dry weight, two parameters that should correlate, showed no significant correlation ($R^2 = 0.1835$). However, when process variability was reduced, the coefficient of determination was over 0.9. The same was true for bulk kilograms and dry weight. The implications are clear. Technical people working in stable, low variation environments are far more able to identify and quantify causal relationships. This is particularly the case where a relationship is suspected, but not yet demonstrated by data. Attempts to demonstrate such relationships are always confounded by instability in the process. Even when a relationship does exist, instability will create a smoke screen that is difficult for technical people to penetrate. Stabilize first is the first principle. Minimize variation around an optimum is the second. There is a place in our world for finesse, but also there is a place for brute force.

PROCESS CONTROL

Recall the Red Beads Experiment. It was a stable process with a process average of 5.7, a UCL of 12.2 and an LCL of 0. Imagine that the experiment were about to be conducted again, using the same equipment and methodology. Is it possible to predict outcomes before the experiment starts?

Providing nothing has changed, it can be predicted that the next set of data will fall at random about the center line, will be bell-shaped, and will remain within the

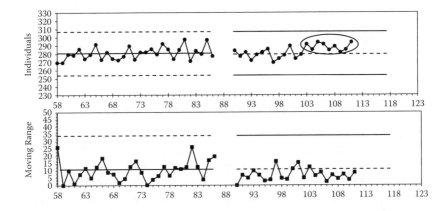

FIGURE 16.3 Example of process control for specific activity.

same control limits. It is impossible to predict what the exact count will be for each sample, but the overall process is quite predictable.

The same applies to the processes in a pharmaceutical plant. Once the process is in a reasonable state of control, the control limits are extended or projected into the future as a prediction of what will happen if nothing changes. Then, as each sample or reading is taken, it is plotted against the projected limits, as seen in Figure 16.3.

Note that when limits are calculated, center lines are solid and control limits are dashed. This pattern is reversed for projected limits so one can tell at a glance whether the chart under examination uses limits calculated from that data or whether they are projected from historical data.

It is absolutely vital that the points are plotted in real time, or in the case where laboratory analysis is required, as close as possible to real time. This is so supervisors, operators, and technical people can see any signal immediately as it occurs. All over the pharmaceutical industry, the end of the month sees all manner of data being analyzed. There is little point waiting until the end of the week or month before asking what caused the out of control point that occurred on the third day of the month. The root causes will likely be lost to antiquity.

For process control, the points are plotted in real time so when a signal is noted, it can be investigated immediately. While the assignable cause is still present, there is a chance of finding it and acting appropriately.

In the example shown in Figure 16.3, a drift up in the process average was noted. This can be investigated and a cause sought so that it can be eliminated (if undesirable) or institutionalized (if desirable). Note that not only is it not necessary to breach the go, no-go, specifications before action is taken, but it also may not be necessary to wait until a point reports beyond the control limits.

The term *statistical process control* (SPC) is applied far too loosely. When one is studying the data to determine stability and to find clues as to how matters might be improved, the activity undertaken is process analysis. Shewhart could have given the charts any number of names. He chose "control charts." It seems likely that the word "control" was supposed to be a clue. Analysis is necessary, but until operators, supervisors, and technical people are using projected limits to control the process minute-by-minute, batch-by-batch, they have process analysis and not process control. Until the process has been stabilized and moved into the process control phase, the activities cannot properly be named SPC.

ELIMINATE WASTE

An oft quoted estimate is that between 20 and 40% of business costs are made up of waste. The pharmaceutical industry is at the high end of that scale. Waste takes many forms. Deviations are a prime example. Long set-up or changeover times are another. Failed batches create huge unnecessary costs. However, perhaps the biggest forms of waste in the pharmaceutical industry are high unit costs and lost gross margin. There is no shortage of examples where pharmaceutical plants have been able to double production once variation was conquered and waste had been stripped out. This slashes unit costs and delays the addition of capital to expand production.

Here is a simple exercise. Follow a batch of product through the process, from start to finish. Measure the amount of time that the product is actually being processed, and how much of the total time is queue, wait, and transit time. A simple analogy may help to clarify terms. Suppose a manager has on his desk an IN and an OUT tray. A clerk drops a file for action into the IN tray. Here it is in a queue. The time spent in the IN tray is queue time. Then the file is placed onto the desk, opened, and actioned. This is processing time. Once processed, the file is placed in the OUT tray. Here it waits to be moved to the chief chemist's desk for further action. This is wait time. Eventually, a clerk picks up the file and carries it to the chief chemist's desk. This is transit time. For some pharmaceutical lines, the total queue, wait, and transit times are greater than the processing time. Set-up time is another area where many plants experience significant waste. The combination of queue, wait, transit, and set-up time blows out cycle time and work in progress. Carrying costs for work in progress can be huge.

Little's law can help here. If variation is reduced throughout the process, cycle time will fall. Then, work in progress will fall, output will rise, or some combination of the two will occur.

If one imagines the process in a perfect state and describes that state, the difference between the imagining and the current reality is waste. The Six Sigma approach quantifies this waste in financial as well as customer terms, to expose it and to act as a spur to remedial action.

Some examples of this waste are the cost of failure and the cost of deviations— not only the failed batches, but also the cost of the deviation process itself. The cost of hiring consultants to help the business pass FDA inspections must be included. To this must be added all the improvement projects technical people and managers might have conducted if only compliance issues did not consume so much of their time. Add to this the cost of retests and additional inspection. Then there is the carrying cost of high work in progress, which is inflated by variation. Finally, add in the lost gross margin mentioned earlier; the cost of lost sales occasioned by low production due to high variability. Where the market for additional product does not exist, instead of lost gross margin, factor in the cost of the labor, supervision, and technical staff that would become unnecessary to meet market demand if only the laboratory and manufacturing processes were operating at Six Sigma levels. All of this is only part of the cost of the hidden factory, the one that produces nothing of value but adds huge cost. Doubtless, readers could add to the list.

Care is required here. Many companies do not figure in the cost of production that should have occurred but did not because of highly variable processes. Even failed batches or products can be costed poorly. Those companies that do calculate the cost of failed product often cost it at manufactured cost, but if there is market demand for this product, the real cost is the lost gross margin. One factory producing medical devices was filling industrial waste bins with scrap and rejects. For their product, demand exceeded supply. The scrap and waste was being costed as manufactured cost. The real losses, measured in lost gross margin, were staggering.

> We don't know what we don't know. We can't address or improve that which we don't know about. Even if we do know we have a problem, we seldom invest to fix the problem if we don't know the total cost.

If the variability in the process is stripped out, along with excessive set-up, queue, wait, and transit time, output soars, unit cost falls, quality improves, deviations plummet, and the carrying costs associated with work in progress falls. Who tracks these hidden costs? Are they in the information and financial systems? The cost of variation is one of the largest costs of doing business, but who has seen it in the monthly financial figures, or in the profit and loss statement?

Here are some rough estimates for the pharmaceutical industry. If Six Sigma levels of performance are achieved throughout a process:

1. Cost reductions of 20 to 40% will be achieved.
2. Capacity will increase by 20 to 50%.
3. The number of employees required will fall by 5 to 20%.
4. Capital requirements will fall by 10 to 40%.

Most amazingly of all, at the outset, there is no need to change anything except how well the people in the process follow the approved procedure with uniformity every time. No change controls are needed in the beginning. Little science or finesse is required. The implementation will be compliant with the Code of Federal Regulations (CFR) and all applicable current Good Manufacturing Practices (cGMPs). To commence, all that is necessary is an understanding of variation, enlightened and determined management, and a measure of brute force. Most managers are surprised at the response from operators after they have begun a determined thrust to minimize variability in everything done in laboratories and plants. A common reaction from the operators is, "And about time, too. We've been saying this needed to be done for years." None of this includes the design of plants and the research and development phases, where the most significant advancements will be made. It is generally accepted that design accounts for 50 to 70% of all potential improvements. In the pharmaceutical industry, this is many billions of dollars.

If the current approaches worked, from commissioning onwards there would be a steady rise in quality and productivity on every production line. Long before the patents expire, every plant would be a compliant, low cost, high quality producer.

In many cases, this is not the situation.

The producers of generic products are successful in part because the big pharmaceutical companies have high levels of waste and unnecessary costs.

Rising healthcare costs are placing ever-increasing pressure on the pharmaceutical industry to bring costs down. Those businesses that take a Six Sigma approach toward understanding and reducing variation are likely to be most effective at eliminating the most significant forms of waste, reducing both unit costs and lost gross margin.

WHAT TO MEASURE AND PLOT

This is an aspect with which many supervisors, managers, and technical people struggle. Often, the problem is that these people think that it is their job to have the answers. It is not. Nobody has all the answers for any process. What may be of use is an approach that leads to the discovery of likely answers.

One under-used element of the DMAIC approach is to assemble a group of people who, as a team, are representative of the entire process. Then brainstorm what are believed to be the key issues and variables under the headings "Inputs, Process, and Outputs." Alternatively, a top down flowchart may be used as the basis for brainstorming.

Using a secret ballot, ask the team members to score the top five variables in each of the three categories, from five (the most important) to one (the least important). Total these scores for all team members and you will have the makings of a Pareto or Juran chart, which will be a good start to identifying the key variables. In nearly every case, this approach will capture most of the key variables, provided:

1. The team is process-based. It should include people from all areas of the process, including analytical and supply. It should include supervision, technical people, and operators. A team made up exclusively of technical people or operators will produce a predictably biased result.
2. Brainstorming is conducted in such a way that no idea is belittled, discussed, or discarded at the outset. They are all written down. A variation on this is to have each team member arrive at the meeting with a list of variables he or she believes are worthy of consideration. The brainstorming then adds to this starting point. The assembled list can then be tidied up by eliminating duplicate ideas.
3. The voting must be by secret ballot. Many people will vote differently if the ballot is public.

The authors have used this approach successfully for many years. There are several derivatives based on the same themes, and readers may choose to use one of these variants. If the team adheres to the above principles, they will discover that there is no shortage of potentially important variables. For a discussion on how and why this approach works, *The Wisdom of Crowds* by James Surowiecki[4] is a good starting point.

REFERENCES

1. W.A. Shewhart, *Statistical Method from the Viewpoint of Quality Control*, The Graduate School of Agriculture, Washington, D.C., 1939.
2. E.L. Grant and R.S. Leavenworth, *Statistical Quality Control*, McGraw-Hill, New York, 1980.
3. D.J. Wheeler, *Advanced Topics in Statistical Process Control*, SPC Press, Knoxville, TN, 1995.
4. J. Surowiecki, *The Wisdom of Crowds*, Abacus, London, 2004.

Appendix A

Example Operational Directive

The Brute Force Approach to Minimizing Variation

Prometheus Biologics
New York
Red River Plant
June 23, 2006
To: Attached Distribution List

OPERATIONAL INSTRUCTION 4/2006

PROJECT TO MINIMIZE VARIATION IN FERMENTATION

SECTION 1 — INTRODUCTION

Our attempts to introduce SPC into manufacturing have been disappointing. Variation remains high and there has been only a small reduction in deviations. It now appears that too much focus was placed on the data or the charts and not enough on the work itself. This directive lays out the next phase in our thrust to achieve stable operations; a project to reduce variability in the fermentation process.

Aim. The aim of this project is to rapidly drive all variation in fermentation to minimal levels. We anticipate that most key characteristics will see significantly reduced variation, that deviations will fall, that capacity will rise, and that causal relationships will become clearer.

SECTION 2 – EXECUTION

General Outline. Every process characteristic will be held to the lowest level of variability possible. This includes: raw materials, sampling, analysis, set-up and operating procedures, the inoculum used, and the maintenance of a repeatable profile for all fermenter control variables including pH, nutrient, and oxygen addition, for every batch. All variables are to be considered, not only those believed to be CTQ. The directive will be executed in three phases:

Phase One. Training and preparation phase.
Phase Two. Processing phase. Thirty batches will be processed under strictly controlled conditions.
Phase Three. Control and analysis phase. Introduce on line SPC. Analysis of the data and recommendations for further activities.

Responsibilities. The following personnel are responsible for conduct of this directive:

Sue-Ellen Schauer, Plant Manager and Project Sponsor
Timothy Stuart, Production Manager for Fermentation and Project Leader
James Samuels, Unit Statistician
Annabelle Corbett, Ph.D., senior unit QA Representative
James Jacobs, David Hughes, Darcy Johnson, and Deedra-Maree Wyborn, Senior
    Process Technicians, appointed as Method Masters
Brian Nunnally, Jr., Analytical Laboratory Manager
Sam White, Consultant

PHASE ONE

Phase one, trial preparation, commenced in May with six activities, all of which will be completed by batch 58:

- The first activities were the one-day seminars for plant personnel conducted by Sam White to provide some introductory training on understanding and reducing variation to Six Sigma levels, to familiarize employees with the impending project, and to solicit the assistance of all in stabilizing operations.
- The second activity was the briefing of individual shift teams by the Method Masters, Timothy Stuart and Sam White, on the critical nature of the elimination of between shift and between operator variability in operating procedures. During these briefings, known differences between shifts were pointed out and handed over to the Method Masters for eradication.
- The Method Masters were released from their normal shift duties on June 5th to allow them to dedicate their time to creating uniform set-up and operating procedures across all shifts, including the preparation of job aids (under guidance from Annabelle Corbett). The Method Masters have each been allocated an area of responsibility within Fermentation. Each is responsible for ensuring that their respective area becomes uniform and remains compliant. They are coordinated by Timothy Stuart and report to him. Timothy Stuart, through the Method Masters, and under advice from

Annabelle Corbett, is the final authority for all operating procedures. Already, these activities have been successful, with some variables, plant upsets, and recent deviation rates showing an improvement. This, in turn, means that data collected during Phase 2 will have minimum corruption due to process variability and plant upsets.

- James Samuels has commenced liaison with Purchasing to minimize the batch changes of raw material for Phase 2. Any unavoidable batch changes are to be logged. He is also to ensure that only inoculum from a single working cell bank are used during Phase 2.
- Brian Nunnally, Jr. has completed work to minimize analytical variability. Laboratory controls are now stable with reduced analytical error. For Phase 2, only two technicians will be used to conduct analyses for critical to quality (CTQ) variables. Brian is also responsible for reviewing sampling methods and assisting the appropriate Method Master with training and job aids to ensure uniform sampling methods are followed.
- James Samuels has commenced work with the control engineers to create a control protocol for Phase 2. Process controllers are to be set to achieve exact repeatability of the batch profile during all Phase 2 batches, and controllers are to engage the process only to protect plant integrity or the safety of personnel. Any adjustments to the process will be made manually and logged.
- Annabelle Corbett is to ensure everything done during the project, including training, methods, and documentation, remains compliant. All activities are to be cleared through her before execution. No changes to GMP are to be made.

PHASE TWO

This phase commences at Batch number 58. Timothy Stuart, the Production Supervisors, and the Method Masters are to maintain a comprehensive log for this phase. Supervisors and Method Masters are responsible for logging and reporting any departure from optimal procedures, regardless of how insignificant these may appear. Every shift is to commence with a planning briefing for the shift and a reminder of the importance of following the job aids exactly.

PHASE THREE

James Samuels will prepare SPC charts for all CTQ variables, with limits projected from Phase 2. Additional training for Method Masters and staff using these charts will be provided commencing from Batch 75. It is imperative that the process be held in as controlled a state as possible following Phase 2. Assisted by Timothy Stuart and selected technical staff, James Samuels will lead in the analysis of the data from Phase 2.

The log created in Phase 2 is to be continued and maintained through Phase 3, until further notice.

Timothy Stuart is responsible for preparing an interim outline report on all three phases within two days of the close of Phase 2. A final report is to be completed within three weeks of the close of phase 2. The final report will include outcomes from the project, recommendations for further improvements, and methods by which the gains made may be institutionalized. In addition, the report is to include recommendations on how any similar activities in the future may be improved.

SECTION THREE - QUESTIONS AND CONCERNS

Personnel who have questions or concerns about the project should direct them to a Method Master, a project team member, or Timothy Stuart, the project leader.

Sue-Ellen Schauer
Plant Manager

17 Improving Laboratories

Revolutionizing Analytical Systems

It seems that production managers are trained to blame the laboratory for any process error. Too often they are correct. The principles described in this book apply equally to the laboratory and can cause transformation.

PRODUCTION LINES ARE THE LABORATORY'S CUSTOMERS

The production unit depends on the analytical laboratory. It deserves the best precision and accuracy possible. It is not possible to judge the quality of the product or to determine the effectiveness of the process without data from the analytical laboratory. With so much riding on the data, it is important for the laboratory to view the production unit as a customer and to deliver the best possible service to this customer. The laboratory processes are no different from any other processes. The laboratory receives raw materials (samples, chemicals), processes the raw materials (sample preparation and analysis), and delivers the final product for use (reports the data). The production unit is expected to produce the highest quality product possible. Should not the laboratory be expected to do the same?

TYPES OF METHODS

All methods are not created equal. They can be broken down into three categories: (1) process relevant tests (PRTs), (2) capability tests (CTs), and (3) residuals tests (RTs). It is important to understand the different test categories to ensure that the proper analysis and control is applied to each type of method.

PRTs are defined as those sets of tests that are used by the manufacturing unit to adjust or control the process. Examples of these tests include potency, quantity, and pH. Errors in PRT measurements can and do wreak havoc in the manufacturing process. The variability associated with these tests needs to be minimized because the analytical variability observed in these tests is transferred directly to the process. This is often the first place to begin the Six Sigma drive to understand and reduce variability within the laboratory.

CTs are those tests whose sole purpose is to determine the quality of the material. Examples of these tests include purity, impurity, and volatiles. Prioritization of CTs depends on the capability of the measurement. (See the next two sections.)

RTs are those tests demonstrating the removal of a material, or are conducted following the disappearance of that material. These tests are characterized by non-detectable or non-reportable values. Often these tests have capabilities in excess of 2.0 or higher. Examples of these tests include host cell proteins, DNA, and solvents. Unless they are incapable, these assays can be prioritized lowest in variability reduction efforts due to their capability and the fact that production does not depend on the data to adjust the process.

VARIABILITY ESTIMATES

Variability comes in three forms: (1) process, (2) sample, and (3) analytical. A control chart of the lot release values will encompass all three forms of variability. It is rare that all three types can be accurately quantitated without a carefully planned and executed hierarchical or similar study. Analytical variability can be determined by validation data, stability data, controls, or preferably, blind controls. Equation 17.1 illustrates the contribution of each form of variation.

$$\sigma^2 = \sigma_P^2 + \sigma_S^2 + \sigma_T^2 \qquad (17.1)$$

where:
σ^2 = Total variance
σ_P^2 = Process variance
σ_S^2 = Sampling variance
σ_T^2 = Test variance

UNDERSTANDING CAPABILITY

The capability (C_{pk}) of a process is a simple parameter to calculate. Equation 17.2 shows the formula. If the capability of a process is 2.0 or higher, the process is considered a Six Sigma capable process and improvement (analytical, sampling, or process) should be prioritized accordingly. Processes that are incapable should be targeted for improvement. If the test method variance is only 10% of the overall variance or lower, the test method is generally accepted as acceptable, but opinions do vary, as do circumstances. The higher the percentage of variability contributed by the method, the higher the priority should be. In general, CTs and RTs will be prioritized lower than PRTs. The production unit does not depend on the data from CTs to adjust the process, and RTs generally have a low priority because of their inherent capability.

$$C_{pk} = \min\left[\frac{USL - \bar{x}}{3\sigma}, \frac{\bar{x} - LSL}{3\sigma}\right] \qquad (17.2)$$

Where: C_{pk} is the capability, *USL* is the upper specification limit, *LSL* is the lower specification limit, $\bar{x}$ is the mean, and σ is the total standard deviation (process, sampling, and test).

Analytical variance may not be a major source of variation, but blind controls or a hierarchical study will soon indicate whether this is the case. In any event, it is necessary to understand the contribution of the laboratory analyses to the overall variation. Usually, the laboratory variability is calculated as a percentage of total variation, to give it perspective.

In too many cases, the laboratory is blamed for sampling variability. A wise laboratory manager will institute statistically valid studies that are effective in quantifying analytical, sampling, and process variability separately. In this way, sound data will indicate where the real problems lie.

ACCURACY VS. PRECISION

Laboratory managers and analysts must understand the difference between accuracy and precision. Too often, these terms are confused. Accuracy issues are attacked in fundamentally different ways than precision problems. Determining whether the source of a problem is accuracy or precision is critical.

The International Conference on Harmonisation of Technical Requirements for Registration of Pharmaceuticals for Human Use (ICH) defines accuracy as "the closeness of agreement between the value which is accepted either as a conventional true value or an accepted reference value and the value found."[1] This is determined by a spike recovery, or orthogonally via another method. Methods utilized in the pharmaceutical industry have a demonstrated accuracy of between 90 and 110% and most are close to 100%. Methods suffering significant matrix effects or other accuracy issues are rare, especially for drug substances and drug products.

Precision is defined by ICH as expressing "the closeness of agreement (degree of scatter) between a series of measurements obtained from multiple sampling of the same homogeneous sample under the prescribed conditions. Precision may be considered at three levels: repeatability, intermediate precision and reproducibility."[1] This is often demonstrated using variance component analysis to isolate the sources of variability in the method, such as analyst-to-analyst, instrument-to-instrument, or day-to-day variability.

With clarity such as this, why then are the terms confused? Figure 17.1 shows a control chart from an assay with excellent precision, but the central tendency or accuracy is approaching chaos. However, without a control chart, how could one be sure that the issue was accuracy rather than precision? Other methods exist, but none of them separate central tendency and variation with such ease and visual impact. The overall σ of this data is 0.315. The chart demonstrates that if the process could be held stable, free of special causes or drifts in the process mean, this σ would fall to 0.136. The reason for the central tendency problem could include either reagent differences or changes to the matrix. The solution to this problem will spring from studies to determine what is causing the mean to shift its location. A good start would be to stratify the data by analyst, instrument, and reagent batch, as discussed in Chapter 16.

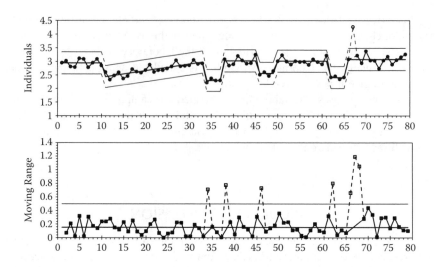

FIGURE 17.1 Laboratory controls — precise but inaccurate.

USE OF VALIDATION DATA TO DETERMINE
LABORATORY PRECISION

The most common method of determining precision in the analytical laboratory is through the use of validation data. Nearly always, this is poor practice. Validations are characterized by the "spotlight effect" (sometimes called the Hawthorne effect). The spotlight effect is a phenomenon where people overestimate the extent to which their actions and appearance are noted by others.[2] The spotlight effect in validation studies refers to underestimation of variability due to intense concentration and the use of highly skilled analysts during special studies such as validations. Analysts selected for validations are usually high performers and even if they are not, they will have an improved focus relative to "normal" samples. Long-term variability or periodicity is not included in validation data due to the short length of time required to collect the data.

Part of the reason Six Sigma calls for such low levels of process variability is to allow a buffer for the inevitable impact of medium- and long-term variation not present in short-term variation, such as will be found with validation data. Less experienced people will replace highly trained analysts, batch changes in reagents will occur, and maintenance issues will arise; these things will happen in even the best-managed laboratories.

Validations are necessary and are an important part of the operation of a pharmaceutical laboratory. Used judiciously, the precision studies for validation help the laboratory to understand the ability of the method to meet the variability requirements and provide guidance in determining the appropriate replicate strategy.

The replicate strategy should include the primary sources of variability, such as analyst-to-analyst and instrument-to-instrument variation. During validation, work

TABLE 17.1
Potency Data for Several Lots

Lot	A	A	A	B	B	B	B	B
Timepoint	1	2	3	1	2	3	4	5
Potency	39.06	42.7	46.53	39.86	40.48	39.79	42.14	37.06

Lot	C	C	C	C	C	C	D	D	D
Timepoint	1	2	3	4	5	6	1	2	3
Potency	36.13	39.36	37.66	38.97	35.2	40.18	40.22	41.34	40.72

to reduce variation in replicates should continue at least until the variability is low enough to meet the customer requirements contained in the protocol.

Validation estimates of the analytical variability are useful, particularly when the method is transferred into the QC laboratory and no other data are available. However, the use of stability, control, or blind control data offers significant advantages to understanding the true variability of the method.

USE OF STABILITY DATA

A simple method for determining the analytical variability of an assay is to use stability data. All stability samples are prepared at the same time and are representative of the production material. This means the sampling variability is usually minimal (although it is not always negligible), and the manufacturing variability is removed. For parameters that do not exhibit degradation or statistically significant change over time, the variability observed in the stability study can provide an estimate of the variability of the analytical assay. This method is an excellent way to estimate variability when control and blind control data do not exist. It can also be a useful population to compare to control data in the absence of blind controls. This estimation is superior to validation data in most circumstances because the spotlight effect has been minimized. However, this estimation can be inferior to both control and blind control data because the data are spread over long periods and they do not reveal day-to-day variability or any periodicity that may be present. Furthermore, if the parameter is degrading, the analysis varies from difficult, provided the fit of the model is excellent and the slopes are consistent, to impossible where they are not.

Pharmaceutical Case Study — Laboratory Precision as Determined by Stability Data

The potency of several lots of a biologic product was measured on stability. The data are shown in Table 17.1. First, the data for each lot were plotted, potency vs. timepoint, and a bivariate (linear) fit was applied. For the data in Table 17.1, none of the bivariate fits were statistically significant (p values were less than 0.05), indicating the parameter did not exhibit degradation. This was corroborated by the technical assessment; the potency was not expected to change over time.

TABLE 17.2
Estimated Method RSD for Potency Data

Lot	A	B	C	D
Mean	42.78	39.88	37.92	40.76
Standard Deviation	3.74	1.83	1.95	0.58
RSD (%)	8.74	4.6	5.13	1.38
Estimated Method RSD (%) = 5%				

This, in turn, indicates the variability observed in the data was due to analytical variation.

The next step was to calculate the mean, standard deviation, and relative standard deviation for each lot. The average standard deviation is the approximate variability for the method. The parameters are shown in Table 17.2. The estimated variability (% RSD) for the method is 5.0%.

USE OF CONTROLS

Controls are one of the simplest ways to understand analytical variability. A good control has several important characteristics. First, the material must be a homogenous, representative material eliminating sampling and process variability. It does not need to come from a single lot, but homogeneity must be assured. Second, the presentation should be identical to the sample. If the sample is a solid, the control should also be a solid. Finally, the material should be stable and should not degrade with time. There are ways to use control charts when this is not true, but it is easier if the material is stable.

Ideally, the control is run in the same way a sample is run. If there are three replicates that are averaged into a single reportable result, then the control should be handled the same way. If this is not done, the control data will lose some utility.

There is some spotlight effect associated with control samples, especially when there are acceptance criteria or system suitability criteria associated with the control. In these cases, most analysts will treat the control in a fundamentally different way than a production sample. This means that the analytical error in production samples is likely to be greater than it is in control samples.

Once the control data is generated, the data should be plotted on a control chart. Prior to release of the data associated with the control, the control chart should be reviewed. It is common practice for the control value to be compared to the control limits only. This is a poor practice. Many signals will be missed if the data are not viewed on a control chart. If the control chart exhibits a special cause, the laboratory result should not be released until a laboratory investigation has been completed.

The following case study shows the use of a control and its ability to determine the precision of an analytical assay.

Pharmaceutical Case Study — Laboratory Precision as Determined by Control Data

Figure 17.2 shows a set of laboratory controls for a pharmaceutical assay. The control chart shows three different populations. The middle population is most intriguing

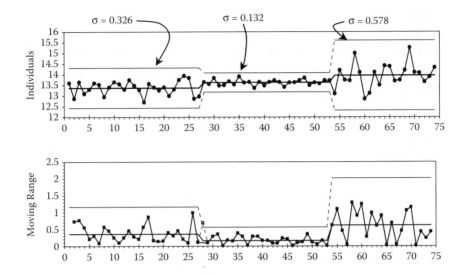

FIGURE 17.2 Unstable laboratory controls.

as this period shows the likely inherent, underlying capability of the method. The analytical variability is shifting over time, greatly increasing in the last third of the control chart.

The standard deviation for each population is marked on the chart, further amplifying the significance of the changes happening over time. A chart displaying several populations indicates when changes happened, providing an opportunity to discover what caused the period of low variation around which laboratory procedures can be standardized to optimize performance.

IMPLEMENTING CONTROLS

Blind control studies are an important approach to understanding the true analytical variability. Running a control along with the process relevant tests is critical. Implementing a control is straightforward.

1. 6 independent points for introductory limits
2. 15 independent points for temporary limits
3. 30 independent points for permanent limits

Since limits can change, it is usually not advisable to put limits into the test method, unless the test method is able to be changed rapidly (e.g., less than one week). Job aides are an excellent place to put control limits, although this may require writing a procedure to control the creation, implementation, and version control of the job aides.

Controls can be characterized, but often this is unnecessary. As stated earlier in the chapter, the control should be:

TABLE 17.3
Comparison of Various Techniques to Determine Analytical Error

Study Type	Short-Term Variability Estimate	Long-Term Variability Estimate	Eliminates the Spotlight Effect
Validation	+	– – –	– – –
Stability data	–	++	–
Control	++	+++	– –
Blind Control	+++	+	+++

1. homogenous
2. representative material
3. presentation identical to the sample
4. stable

If the control meets these criteria, characterization should not be necessary. The focus of a characterization study, if performed, should be to determine if the control criteria are met and to establish introductory limits.

BLIND CONTROLS

The use of blind controls is rare in the pharmaceutical industry. The roots of this probably lie in the compliance aspects of the study (covered in Appendix B to this chapter). However, there is no better way to understand the true variability of the analytical method. This is because the analyst treats the material like any other sample. The key to a blind control study is to keep the group organizing the study as small as possible and ensure the study is kept secret from everyone else in the organization. The analyst must not be able to tell the difference between a production sample and a blind control.

Table 17.3 gives an indication of why blind controls are a superior way to determine the true analytical variability, as it will be present in production samples. The only limitation associated with a blind control study is associated with the long-term variability estimate. This limitation can be overcome with a long enough series of blind controls.

Controls are part of the normal protocol for most laboratories. They are nearly always available. Table 17.3 illustrates that if these are supplemented with occasional blind controls, a very complete picture of the analytical variability is exposed.

Pharmaceutical Case Study — Blind Control Study

In this example concerning tests for a consumer healthcare product, customers were dissatisfied with cycle times. In their own defense, the laboratory cited high workloads, which sometimes led to a backlog of tests to be conducted. The plant had experienced regulatory problems to the extent that it was operating under a consent decree. The laboratory had doubled in size to accommodate the additional testing required.

A study of the laboratory processes was undertaken. Retests and laboratory deviations were quite high, and these events were consuming a considerable amount of laboratory time. Batches were failing final test and becoming quarantined. The subsequent testing and investigation almost invariably released the batch, but the cost in terms of work in progress and laboratory time was high.

Cycle times for this test were studied. They revealed that the work did not flow through the laboratory evenly. Rather, it could be highly erratic. Sometimes, tests were completed promptly. At other times, significant delays were incurred.

Then the test error, which was claimed to be ±1.5%, was examined. The laboratory manager was asked to participate in a blind control study. He refused, stating that he already knew the test error.

The production manager and a consultant appointed by the vice president of manufacturing went ahead with the blind controls without the laboratory manager's knowledge. A vat of finished product was agitated thoroughly to ensure it was homogenous and the product was decanted into bottles. Approximately 50 of these bottles were sent to the laboratory mixed in with the production samples. Sample numbers of the blind controls were recorded to enable them to be separated from the routine production samples.

The analytical results for these samples displayed variation of approximately ± 3%. It transpired that ±1.5% was the instrument error only. It did not take into account the between-instruments and between-analysts variation, which was likely to have increased commensurate with the increase in analysts. The hiring of inexperienced analysts during the expansion of the laboratory is likely to have added to analytical error.

The original estimate of a test error of ±1.5% did not take into account other aspects such as environmental control or batch changes for reagents and other chemicals used in the laboratory. Nor did it include maintenance or calibration of instruments. The blind controls revealed all of this additional variation. Given that the specifications were ±5%, it soon became apparent why batches were failing final inspection. In addition, the control chart of the laboratory controls in Figure 17.3 demonstrated that the analytical error was unstable.

The jump in variation noted in the chart occurred at the time the laboratory was expanded rapidly to accommodate the additional testing required by the consent decree. This increased analytical error was the most significant factor in the erratic cycle time. Every quarantined batch, deviation, and investigation reduced the laboratory's capacity to run tests. This, in turn, slowed the cycle time and made it erratic. The true analytical variability of the test would not have been known without the use of a blind control. Any other measure, be it validation or stability data, probably would not have shown the variability the way the blind control did. The normal laboratory controls did not. For the period up to control number 29, the analytical error of the first four populations, if each is taken independently, was ±1.4% (calculated from $\bar{R}$). If these four populations are grouped together, the total analytical variability is ±2.7%.

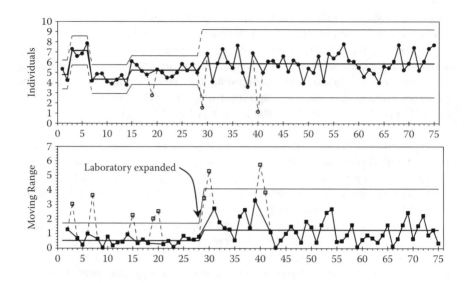

FIGURE 17.3 Highly variable laboratory controls.

REDUCING VARIABILITY — MORE IS NOT ALWAYS BETTER

The goal of all laboratory managers should be to continuously improve the quality of the data produced by their laboratory. This means understanding and reducing the analytical variation. As was discussed earlier, the first focus should be on the PRTs and incapable CTs. Once the customer valid requirements for these methods are satisfied, the ability exists to prioritize which methods will be focused on next. Should the variability in PRTs be further reduced? Should the next target be the incapable CTs? Do any RTs exist that would benefit from variability reduction techniques? Six Sigma methodologies can help here. A simple DMAIC process can be used to determine the prioritization. This is a nice problem to have. Once PRTs are no longer problematic, the CTs are all capable, and the RTs are all performing well, the laboratory is in a position to choose where next to reduce variation below the customer valid requirements, and it should do exactly that.

Occasionally, the customer valid requirements will change, or a relative standard deviation (RSD) of 10% that was satisfactory last week may not be acceptable next week. A common response is to test more samples. The following example demonstrates why testing more samples is not always the correct solution.

Pharmaceutical Examples

Replicate strategy. A laboratory manager queried a development colleague on the reason for using a triplicate injection approach for an HPLC assay. The laboratory manager asked the scientist to show the data on the injection-to-injection variability that warranted a triplicate injection approach. The scientist told the laboratory manager that three injections were needed because "it was more than one injection." There is an old tongue-in-cheek laboratory axiom, "An n (sample size) of 1 is random, an n of 2 is coincidence, and an n of 3 is science." Here was a major investment of resources that resulted in no substantial increase in the quality of the data because the variability has not been significantly altered by the replicate strategy. Moreover, the replicate strategy was not based on an attempt to improve the data. It was based on the notion that three is bigger, and therefore better, than one.

Changed customer valid requirements. When the customer valid requirements change, the first step is to determine whether the existing method is capable of meeting the new requirements. If control or blind control data exist, they can be used to modify the replicate strategy in an attempt to demonstrate if the new requirements can be met. Additionally, a DOE study created to understand the sources of variability (variance component analysis) can be used to determine the appropriate replicate strategy. If homogeneity is an issue, more samples will be needed, but testing additional samples does not always significantly reduce the analytical variability. Ultimately, the replicate strategy should be designed to result in a minimization of variability, but to what extent does a laboratory manager pursue reduced analytical error? A rule of thumb in the pharmaceutical industry is for the analytical variability to be no more than 10% of the total variability (process, sampling, and analytical). This is not an absolute, nor is it always achievable. It is a guide. It is not a universally accepted standard. Finally, the customers of the laboratory deserve the best analytical precision possible, especially for PRTs. The history of all analytical techniques is an ongoing demand for increased precision. The laboratory can wait until it is forced to reduce variation by customer requirements, or it can get ahead of the game.

PHARMACEUTICAL CASE STUDY — REDUCTION OF VARIABILITY

The degree to which production managers are dependent on the data produced by the analytical laboratory is profound. If the laboratory sneezes, the production department contracts influenza. In this case study, the production managers were frustrated by the perceived inaccuracy of the data being produced by the analytical laboratory. A meeting was called to discuss the issues and propose a solution. The production managers favored taking additional samples. The laboratory manager recognized there were no homogeneity issues and the production managers confirmed this; therefore, additional samples would not solve the problem.

The laboratory manager correctly assessed the problem was not one of accuracy, but of precision. The analytical variability for the assay was too large, creating the illusion of accuracy issues. The laboratory manager embarked on a variability reduction project (see guidelines in Appendix A at the end of this chapter). The project focused on reducing the analytical variability for each test by reducing the controllable variability such as analyst-to-analyst differences and standardizing the practices within the laboratory.

The results of the laboratory manager's effort were as profound as they were immediate, as noted in Figure 17.4. The data prior to and after the change were stable, with the exception of one special cause point. The three sigma limits after

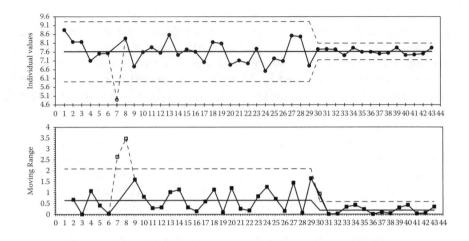

FIGURE 17.4 Laboratory controls showing variability reduction.

the variability reduction program were less than one third of the original limits. The average of the data did not change after the implementation of the variability reduction program, which demonstrated the correctness of the laboratory manager's assessment that the issue was one of precision.

During the presentation of this case study to other laboratory managers and technical personnel, many were interested in knowing which critical parameters had been addressed to result in this improvement. This is a dangerous question because it suggests that these laboratory managers were interested in examining the so-called "critical parameters" only.

Most "critical parameters" are not critical to the variability at all; they may be critical to accuracy. Too often, the studies to demonstrate that these parameters are truly "critical" are not statistically powered to provide adequate information about precision. Critical parameters are often the result of information passed down from scientist to scientist like a campfire story. While there are examples where the critical parameters have been adequately identified, there are many where they have not.

It is certainly helpful to understand the source of the variability improvement. When variability is as low as noted in Figure 17.4, it is likely the source will be found. However, the purpose was to reduce variation, all controllable variation, not just some of it or even most of it. To focus only on the critical parameters would mean giving up the opportunity to reduce variability to a minimum, and would force production to receive data that are more variable than they need to be. This is not acceptable in a Six Sigma environment. It is a fair bet that laboratory managers claim all possible deductions when filing their tax returns. A similar approach is called for in a laboratory. All possible variability deductions should be made, not only those that happen to be in an area of interest. The customer deserves nothing less.

Variability reduction in a laboratory could be couched as a formal Six Sigma project, but should not maximizing precision be a normal part of a laboratory manager's job? Must they wait until a Six Sigma Black Belt gives them permission to improve, or for the launch of a formal project? Laboratory personnel are highly trained and already have most, if not all, of the tools and techniques necessary to improve precision. First, stabilize the system. Reduce the controllable variation to a minimum. If needed, conduct DOE studies to isolate rogue variables and implement process improvements to further improve both precision and accuracy. Most laboratory managers already have the skills required to do these things. So, what is stopping them?

IF STANDARDS ARE MET, WHY BOTHER REDUCING VARIATION?

Many laboratory managers are reluctant to drive toward Six Sigma levels of performance. This is especially true when existing standards are being met, at least for the most part. This attitude is a function of the old go, no-go mentality so common in the industry. There are several reasons why minimizing analytical variation should not be an option. First, the people who need pharmaceutical products deserve the best quality possible. Second, the production people deserve the best analyses possible. Analytical error adds to process variation; it helps to shroud causal relationships as well as small signals in the production process. Third, the lower the analytical variability, the fewer will be the number of deviations that must be raised and cleared. Laboratories need not become better and faster at clearing deviations. They need to eliminate them as far as possible. Finally, the laboratory managers who earn a reputation for Six Sigma levels of precision as well as excellent customer service do their career prospects no damage at all.

There are no good reasons for strict adherence to the old go, no-go mentality so prevalent in the pharmaceutical industry, but many good ones that support the adoption of Shewhart's methods and a Six Sigma approach to laboratory precision. It can be done. It has been done. It will be done again.

ONE POINT LEARNING

1. Stabilize first, and then minimize variation around an optimum. Close enough is not good enough, nor is merely meeting standards.
2. Understanding the difference between accuracy and precision is vital. The control chart is an excellent tool for this purpose.
3. Allocate priorities; start with PRTs.
4. Blind controls are superior at demonstrating analytical variation.
5. More samples and tests are not always a satisfactory approach. Often, they add costs and little or no value.
6. Understanding and reducing variation is the job of everyone in the laboratory.

REFERENCES

1. ICH Q2(R1): Validation of Analytical Procedures Text and Methodology. International Conference on Harmonisation Guidelines, www.ich.org.
2. T. Gilovich, V.H. Medvec, and K. Savitsky, The spotlight effect in social judgment: An egocentric bias in estimates of the salience of one's own actions and appearance, *J. Pers. Soc. Psychol.*, 78(2), 211–222, 2000.

Appendix A

Implementing a Laboratory Variability Reduction Project

Since the procedures used do not need to be modified, the project can be implemented within one day. It is recommended that the people involved receive some training on the importance of understanding and reducing variation before project launch. Once the method is selected, the analysts running the method should be brought together to discuss how the method is run. This discussion should focus on ambiguities in the method that involve analyst interpretation, along with areas where ranges are provided. An analyst, rather than a manager, should be identified to lead the project. An analyst will have more credibility in these matters, is closer to the work, and will better ensure all factors are considered. Experience demonstrates that senior analysts are more demanding of their peers in the pursuit of precision than are managers. Management can ensure that the groups have full support and their needs are met to accomplish their mission, but the analysts, with particular regard to the project leader, will determine the success or failure of a project like this.

The analysts discuss the various ambiguities and ranges and decide how the method will be run. No change control is needed to implement the project; the method does not need to be modified or rewritten. The intent of the project is to ensure rigid uniformity among the analysts. Once the decision is made as to how the method will be run, the analysts can begin running the method in this fashion. A job aide can be written to explain how the parameters will be followed to ensure the uniformity can be passed along to new analysts and existing analysts can remember what was decided. Any small variation observed should be noted and the result flagged. This will help later investigation into the effect of the change. Parameters that should be considered include: pipetting technique (e.g., positive displacement, pre-wetting.), dilution schemes (e.g., how do you do a 1:10 dilution, dilution limit), and incubation times (e.g., fix a specific time). There are countless opportunities or "events" (recall the manufacture of a cell phone), even with simple methods. The purpose of the exercise is to eliminate, as much as possible, the analyst-to-analyst contribution to the overall variability.

The potential is significant. For one potency method, the group showed a 50% reduction in variability. For another measurement, the reduction in variability was over 75%. The exciting aspect of this approach is that the project risks nothing. There is nothing to lose except analytical error.

Appendix B

Implementing a Blind Control Study

The key to a blind control study is to ensure the laboratory analyst cannot tell the difference between a production sample and a blind control. Keep the group organizing the study as small as possible and ensure the study is kept secret from everyone in the organization except the small group planning and executing the study. A laboratory representative, production representative, and a QA representative can implement the study in most companies.

The most difficult aspect of a blind control study is getting the samples *into* the laboratory system. Depending on the sample submission system and the overall workload, this can vary from simple to extremely difficult. A discreet, knowledgeable laboratory representative can assist with this aspect without violating the compliance rules of the laboratory. The laboratory representative is responsible for discretion, assisting with sample submission, monitoring the study, and dealing with issues within the laboratory. The production representative is responsible for discretion, preparing the controls, and submitting the samples. The QA representative is responsible for discretion, ensuring the compliance aspects of the study are addressed, and dealing with any quality-related issues that arise, such as laboratory deviations associated with the samples.

The first step is to write a protocol. The protocol should be concise, covering the areas needed. An example protocol is shown in Figure 17.5. The protocol is designed to explain the purpose of the analysis and provide documentation of agreement to the parameters and plans for the study. Minimally, the protocol should state the purpose of the study, the samples to be used for testing, the method being studied, any acceptance criteria or important compliance parameters, and approvals. The next steps are to acquire the needed materials and to implement the protocol. Of course, a technical report should be authored once the study is complete. The final step is to present and discuss the data with all interested parties. Care should be taken to focus on the data and to avoid using it in a punitive way.

Title: Blind control study for Test Method TM-1234, Potency of BSA drug substance

Author: Brian K. Nunnally, Jr, QC Analyst

Purpose: In order to understand and reduce the variability of TM-1234, Potency of BSA drug substance, a blind control study is being undertaken in QC.

Experimental: Approximately 1L Samples will be taken from BSA batches HI-HOAG, FE-OZZI, and SN-CANN and mixed. The mixture will be aliquoted into 100 10mL vials. Vials will be labelled with sequential lot numbers beginning with AU-FNGR.
Samples will be submitted into the laboratory by Timothy Stuart, Production Supervisor randomly, up to 5 times a week, along with the laboratory paperwork associated with the sample. Data will be reported into the Lims system.

Method: TM-1234

Acceptance Criteria: Results generated using method TM-1234 will meet its system suitability criteria (R^2 of standard curve and slope of standard curve). The data generated will be analysed to determine overall variability and variance components. The study will end if: 1) the authors and approvers agree sufficient data has been collected, 2) the blind control study is compromised, or 3) no additional samples are available. Any deviations associated with the samples will be resolved by the team and cleared in a way ensuring the study is not compromised.

Approved:
Timothy Stuart, Production Supervisor
Danielle Corbett, Ph.D. QA Representative

FIGURE 17.5 Example blind control study protocol for a fictional BSA potency assay.

18 Beyond Compliance

Remaining Compliant — Necessary but Insufficient

> The entire history of pharmaceuticals is an ever-increasing demand for reduced varia-
> tion. We can be dragged there by tightened industry and FDA requirements, or we can
> get ahead of the game while improving profits.

An error is made in the laboratory. The analyst used a citrate buffer instead of the phosphate buffer specified by the procedure, a Good Manufacturing Practice (GMP) violation. It happens. Later that day, an out of specification (OOS) release result surprises everyone. The batch is quarantined and the procedures are checked to determine the sampling regime required to determine whether the OOS was a rogue test result or a genuine process problem. A few hours later, while the plant manager sleeps, an operator messes up the sequence for a batch set-up. The batch is scrapped and the shift supervisor checks her documentation to ensure the inevitable investigation is compliant, even if the batch is not.

Deviations are raised. Investigations are conducted. The second assay, with a phosphate buffer this time, gives a good result. The quarantined batch is cleared for release when the investigation determines an analytical error was the real issue. On another line, a new batch is rushed into production after the scrapped batch has been cleared from the process. In the fullness of time, all the deviations are cleared. The relevant processes remain compliant and medicine flows into the marketplace.

The authors have been telling this story at seminars for the pharmaceutical industry for many years. The audience, for the most part, is puzzled. It all seems so routine. Deviations happen. They are cleared and life goes on, as does production. What, they ask themselves, is the point of the story?

Deming was fond of telling a story about a student he would pick up on his way to lectures. The student had a young daughter. One day Deming turned up only to discover that his passenger was running late. The daughter invited him inside and offered Deming coffee and toast.

Deming thought that sounded fine, but was unsure whether the girl was old enough to make toast. He said as much to the girl, who was deeply offended. Toast? Of course she knew how to make toast! She'd been watching her mother do it for years. It was simple. One placed bread in the toaster. Once the toaster had popped, one took the toast to the sink and scraped off the burnt areas.

The girl could imagine no other way to make toast. This method was her uniform experience. She could scarcely imagine a world where toast was never burnt. It was beyond the realm of her understanding.

In a similar way, many people in the pharmaceutical industry do not understand the moral behind the opening story to this chapter. In a like manner to the girl, they can scarcely imagine a world where deviations rarely happen. Their world is one where the business remains compliant by raising and clearing deviations. Remaining compliant is chief among issues in a strictly regulated world governed by a go, no-go mindset; but it is not necessary to clear deviations to remain compliant. Compliance could be achieved by eliminating deviations as much as possible.

The moral of the opening story is that every month the pharmaceutical industry spends millions of dollars scraping burnt toast. Perhaps it is time the industry decided to fix, adjust, or replace the toaster.

It ain't so much the things we don't know that get us in trouble. It's the things we know that ain't so.

Artemus Ward
1834–1867

In the pharmaceutical industry, a major focus is placed on remaining compliant, sometimes to the exclusion of all else. Compliance is essential; there is no doubt about that. However, when people from outside the industry enter it and find thousands of people working to raise, investigate, and clear deviations, they are at a loss to understand what is happening. The experienced people from the pharmaceutical industry explain the critical nature of remaining compliant, complete with a comprehensive set of war stories to illustrate their dilemma. The outsiders wonder why the pharmaceutical people don't stabilize the processes and eliminate as many deviations as possible.

When outsiders and newcomers to the industry urge a drive to fix the processes, often the response is the kind of knowing smile usually reserved for simpletons. It is explained that changing the GMP process is simply not an option, and that continuing to raise and clear deviations is the only viable option. Change agents in the industry face pre-existing momentum of battleship proportions.

Essentially, there are two ways to remain compliant. The first is to capture the errors and omissions before they reach the marketplace, to investigate them to ensure that neither the product nor the consumer will be compromised, and to take some form of corrective action. The second is to create such brutal uniformity that deviations of any kind are rare. In accordance with the Pareto principle or, more properly, with Juran's rule, approximately 85% of the deviations arise from approximately 15% of the causes. The job is nowhere near as big as it seems. In previous chapters, we provided examples of massive improvements in uniformity. In one laboratory, deviations dropped to less than half their previous level in three months.

The cost of compliance is staggering. One only needs visit pharmaceutical facilities a few times before one is likely to encounter a plant where employees are outnumbered by former FDA personnel and consultants hired to help the plant pass an inspection. The cost is vast. Even this cost pales in comparison with the cost of lost production and, therefore, lost gross margin. Pharmaceutical plants with the capacity to double production are not rare.

If the current approaches worked, from commissioning onward, there would be a steady decline in deviations on every production line, and productivity would rise with time.

In most cases, it does not.

Working harder using existing approaches will achieve little.

A new operating philosophy and methodology are required.

Some compliance costs are fixed. These include the minimum FDA required testing to guarantee the process and the product. An interesting exercise is to imagine stable laboratory and manufacturing processes operating at Six Sigma levels of performance. Then calculate the minimum compliance costs for such a situation and compare it with the actual costs. The difference between the two figures will be large. Included in these compliance costs must be the cost of failed product, additional inspection and testing, unnecessary staffing levels, lowered production, and lost gross margin that can be traced to deviations and the variation they introduce.

It is axiomatic that deviations are caused by variation. The analyst varies from the procedure in selecting a buffer. An operator varies from the standard but perhaps ambiguous GMP set-up procedure. There is minute-to-minute and batch-to-batch variation in every process. Stable systems are uncommon in the pharmaceutical industry, but it is worth recalling that for managers bent on improving matters, chaos is a blessing.

If the methodology used by so much of the pharmaceutical industry actually worked, any plant should see a decline in the total number of deviations as the months and years pass. In many cases, they do not. The chart in Figure 18.1 comes from a pharmaceutical company. Only deviations from plants and laboratories in operation for the entire four years covered by the chart are included. Rarely does

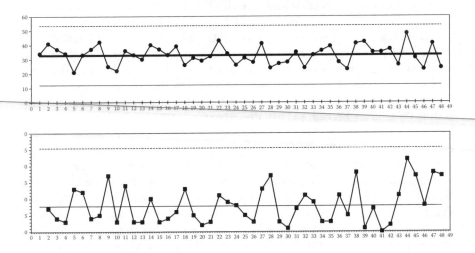

FIGURE 18.1 Control chart for deviations.

one find such robust stability. This chart tells the managers from this company that whatever they are doing to attend to the root causes of deviations is not working. Unless they are prepared to take a different approach, the chart will continue to exhibit stability. Working harder at old methodologies will not help. Transformation is required; that is, a metamorphosis.

In a period of three months, one consultant visited a number of pharmaceutical plants in North America and Europe. In the first U.S. plant visited, his statistical analysis revealed that a large proportion of deviations could be traced to poor set-up and start-up procedures. After further investigations with supervisors and operators, he recommended some form of check sheet to help operators ensure the correct start-up procedure was followed every time. The staff in this plant went to great lengths to explain how such documents were not included in the existing GMP documentation and would be illegal unless the GMP documents were rewritten and revalidated, something all were keen to avoid. This is a design issue. At the next U.S. plant visited, the staff showed the consultant what they called patrol lists and job aides that were exactly what had been recommended at the previous plant. Their hope was that the consultant could help them improve these documents, as well as their batch set-up and operating procedures.

What had been found in the first plant had little to do with GMP and much to do with avoidance and denial. The remedy for compliance issues is less likely to be technical in nature than it is to be operational. Such operational issues require determined leaders who are willing to do that which is necessary to make the process work every time.

The human cost of non-compliance is worth considering. During the same tour of plants in North America and Europe, the consultant found young scientists bowed under the weight of deviations and compliance issues. They trained as scientists, but they found themselves consumed by raising and clearing deviations that should never have happened. Their frustration was such that many of them have been lost to the industry. One young European scientist remarked, "If this is the way the pharmaceutical industry practices chemistry, I'm removing the golden handcuffs before they become too heavy." He did, and another bright, energetic young scientist left the industry.

Generally speaking, large corporations in any industry manage to expel mavericks and young Turks who urge profound change. One pharmaceutical industry executive described them as "hair shirts" because of the discomfort they caused. Nevertheless, he was certain that only the mavericks could bring about the transformation the industry needed, and that the industry needed to nurture and promote its mavericks, rather than to drive them away.

The pharmaceutical industry goes to great lengths to people its corporations with highly qualified technical staff. One visitor to the industry commented that it had more Ph.D.s per acre than any other industry he had encountered. Most of the core manufacturing and compliance problems in the pharmaceutical industry are not technical in nature. They are operational. Nevertheless, the tendency is to address these problems from a technical perspective because that is the nature of the people doing the work. This surprises no one, given that technical excellence is a prime consideration for advancement in the industry. What surprises some is the lack of training in the management of complex operations provided to graduates working in the industry.

Despite the trouble and expense the industry incurs to staff itself with talented technical people, in one pharmaceutical corporation not one operations research graduate could be found. Perhaps this explains why almost no one had heard of Little's Law, and little was being done to apply this kind of thinking to improve processes. Plant managers are more often appointed based on their technical skills than they are on their ability to manage a factory. However, being an outstanding chemist or biologist is insufficient qualification to become a plant or manufacturing manager. Such positions call for an understanding of manufacturing theory and practice; an understanding of how to conquer variation in manufacturing and laboratory processes, together with the determination to do so. The industry is faced not with replacing or upgrading technical skills, but with adding to them.

> The long range contribution of statistics depends not so much upon getting a lot of highly trained statisticians into industry as it does in creating a statistically minded generation of physicists, chemists, engineers and others who will in any way have a hand in developing and directing the production processes of tomorrow.

> **W. A. Shewhart**

WE HAVE MET THE ENEMY, AND HE IS US

Walt Kelly first used the quote "We have met the enemy, and he is us" on a poster he created for Earth Day in 1970. The poster shows a forest choked with rubbish and makes the point that people everywhere had no one but themselves to blame for their pollution problems. The poster is analogous of any industry. The good news is that the pharmaceutical industry's problems belong to the industry, not the FDA, nor the regulations that bind the industry. Once those of us who work in the industry accept that we own the problem, the power to remedy the situation lies in our own hands. We can choose to maintain the existing reality, or we can choose to create a new reality; one characterized by brutal uniformity in everything we do.

We have the knowledge and the power to move beyond being satisfied with remaining compliant by raising and clearing deviations. We can all but guarantee compliance by driving manufacturing and laboratory processes to Six Sigma levels of performance. In this environment, when deviations do occur, they surprise everyone. The choice is ours.

Appendix 1

Factors for Estimating σ' from $\bar{R}$ and $\bar{\sigma}$

Sub-Group Size n	Factor d_2 $d_2 = \dfrac{\bar{R}}{\sigma'}$	Factor C_2 $C_2 = \dfrac{\bar{\sigma}}{\sigma'}$
2	1.128	0.5642
3	1.693	0.7236
4	2.059	0.7979
5	2.326	0.8407
6	2.534	0.8686
7	2.704	0.8882
8	2.847	0.9027
9	2.970	0.9139
10	3.078	0.9227
11	3.173	0.9300
12	3.258	0.9359
13	3.336	0.9410
14	3.407	0.9453
15	3.472	0.9490
16	3.532	0.9523
17	3.588	0.9551
18	3.640	0.9576
19	3.689	0.9599
20	3.735	0.9619
21	3.778	0.9638
22	3.819	0.9655
23	3.858	0.9670
24	3.895	0.9684
25	3.931	0.9696
30	4.086	0.9748
35	4.213	0.9784
40	4.322	0.9811
45	4.415	0.9832
50	4.498	0.9849

Note: All factors assume normality.

From values tabled by E.L. Grant and R.S. Leavenworth, *Statistical Quality Control*, McGraw-Hill, New York, 1980. With permission.

Appendix 2

Factors for $\bar{x}$ and R Control Charts

Sub-Group Size n	$\bar{x}$ Chart Factor A_2	Ranges Chart LCL Factor D_3	Ranges Chart UCL Factor D_4
2	1.88	0	3.27
3	1.02	0	2.57
4	0.73	0	2.28
5	0.58	0	2.11
6	0.48	0	2.00
7	0.42	0.08	1.92
8	0.37	0.14	1.86
9	0.34	0.18	1.82
10	0.31	0.22	1.78
11	0.29	0.26	1.74
12	0.27	0.28	1.72
13	0.25	0.31	1.69
14	0.24	0.33	1.67
15	0.22	0.35	1.65
16	0.21	0.36	1.64
17	0.20	0.38	1.62
18	0.19	0.39	1.61
19	0.19	0.40	1.60
20	0.18	0.41	1.59

For individual point and moving range charts: $A_2 = 2.66$, $D_3 = 0$, $D_4 = 3.27$.

All factors assume normality.

From values tabled by E.L. Grant and R.S. Leavenworth, *Statistical Quality Control,* McGraw-Hill, New York, 1980. With permission.

Appendix 3

Factors for $\bar{x}$ and σ Control Charts

Sub-Group Size n	$\bar{x}$ Chart Factor A_1	σ Chart LCL Factor B_3	σ Chart UCL Factor B_4
2	3.76	0	3.27
3	2.39	0	2.57
4	1.88	0	2.27
5	1.60	0	2.09
6	1.41	0.03	1.97
7	1.28	0.12	1.88
8	1.17	0.19	1.81
9	1.09	0.24	1.76
10	1.03	0.28	1.72
11	0.97	0.32	1.68
12	0.93	0.35	1.65
13	0.88	0.38	1.62
14	0.85	0.41	1.59
15	0.82	0.43	1.57
16	0.79	0.45	1.55
17	0.76	0.47	1.53
18	0.74	0.48	1.52
19	0.72	0.50	1.50
20	0.70	0.51	1.49
21	0.68	0.52	1.48
22	0.66	0.53	1.47
23	0.65	0.54	1.46
24	0.63	0.55	1.45
25	0.62	0.56	1.44
30	0.56	0.60	1.40
35	0.52	0.63	1.37
40	0.48	0.66	1.34
45	0.45	0.68	1.32
50	0.43	0.70	1.30

All factors assume normality.

From values tabled by E.L. Grant and R.S. Leavenworth, *Statistical Quality Control*, McGraw-Hill, New York, 1980. With permission.

Index

A

A$_1$ factor, 138, 195
A$_2$ factor, 139, 193
Accountability, 23
Accuracy, 96, 169
Analytical variability reduction
  accuracy vs. precision, 169
  batch potency control example, 91–92
  blind control studies, 169, 173–175, 183–184
  controls, 172–175
  critical parameters and, 178
  customer valid requirements, 177
  determining precision, 170–172
  "deviations happen" thinking, 185
  pharmaceutical examples, 176–178
  prioritization, 176
  process capability and, 168–169
  project implementation issues, 181
  resistance to change, 5–6, 179
  sample operational directive, 165
  stability data, 171–172
  types of tests, 167–168
  validation data, 170–171
  variability estimates, 168
Asimov, Isaac, 102
Assembly costs, component quality improvement and, 15–16
Assignable causes, *See* Special causes; Unstable variation
Attributes charts, 105–106, 139
Auditors, 79
Automated production lines, low tolerance for variation, 86–87
Automatic process control
  over-control issues, 61–65, 130
  serial sampling and sub-group integrity, 122–123
  testing with trials, 64–65, 84, 130
  unnecessary calibration, 65
Average (ξ), 30–31, *See also* Average and range control charts
  average of, 112
  calculating for average and range control chart, 111–112
  normalization of, 130
  range and, 106
  rms standard deviation calculation, 36

shifts in control chart means, *See* Shifts in process mean
standard error of the mean, 129
Average and range control charts, 111, *See also* Control charts
  averages vs. individual values, 130
  constructing, 111–115
  deviations from normality and, 130, 132
  factors for, 193
  interpreting, 131–134
  investigation guidelines, 133–134
  origins of formulae, 127–128, 137–140
  over-controlled process example, 118–119
  range average interpretation, 115–118
  sampling modes and, 119, 121
  sensitivity to shifts in the process mean, 130
  Shewhart's bowl experiment, 125–127
  small sub-groups of data, 111–112
  special causes and, 117
  specifications or tolerance and, 129
  sub-group integrity, 119–123, 152–153
  sub-group size effects, 128–129
  tests for stability, 126, 132–133
  why chart works, 118–119
Average and sigma control charts, 127–128, 137–139
  factors for, 195

B

B$_3$ factor, 139, 195
B$_4$ factor, 139, 195
Background noise, 141
Batch processes
  mass production principles, 28
  steady-state trial, 83–85
  sub-batch potency testing, 89–92
Bell-shaped distribution, *See* Normal (or bell-shaped) distributions
Bell Telephone Laboratories, 39
Black belts, 22
Blessing of chaos, 153
Blind control studies, 169, 173–175, 183–184
Bottling best practices, 87
Box, George, 34
Brainstorming, 160
Brute force approach, 153–155, 163–166
Business strategy, 21

197